THRESHOLD LOGIC

ELECTRICAL SCIENCE

A Series of Monographs and Texts

Edited by

Henry G. Booker
UNIVERSITY OF CALIFORNIA AT SAN DIEGO
LA JOLLA, CALIFORNIA

Nicholas DeClaris
UNIVERSITY OF MARYLAND
COLLEGE PARK, MARYLAND

JOSEPH E. ROWE. Nonlinear Electron-Wave Interaction Phenomena. 1965

MAX J. O. STRUTT. Semiconductor Devices: Volume I. Semiconductors and Semiconductor Diodes. 1966

AUSTIN BLAQUIERE. Nonlinear System Analysis. 1966

VICTOR RUMSEY. Frequency Independent Antennas. 1966

CHARLES K. BIRDSALL AND WILLIAM B. BRIDGES. Electron Dynamics of Diode Regions. 1966

A. D. KUZ'MIN AND A. E. SALOMONOVICH. Radioastronomical Methods of Antenna Measurements. 1966

CHARLES COOK AND MARVIN BERNFELD. Radar Signals: An Introduction to Theory and Application. 1967

J. W. CRISPIN, JR., AND K. M. SIEGEL (eds.). Methods of Radar Cross Section Analysis. 1968

GIUSEPPE BIORCI (ed.). Network and Switching Theory. 1968

ERNEST C. OKRESS (ed.). Microwave Power Engineering:
Volume 1. Generation, Transmission, Rectification. 1968
Volume 2. Applications. 1968

T. R. BASHKOW (ed.). Engineering Applications of Digital Computers. 1968

R. LYON-CAEN. Diodes, Transistors, and Integrated Circuits for Switching Systems.

M. RONALD WOHLERS. Lumped and Distributed Passive Networks.

MICHEL CUENOD AND ALLEN E. DURLING. A Discrete-Time Approach for System Analysis.

In Preparation

JULIUS T. TOU (ed.). Applied Automata Theory.

HUGO MESSERLE. Energy Conversion Statics.

GEORGE TYRAS. Radiation and Propagation of Electromagnetic Waves.

THRESHOLD LOGIC

CHING LAI SHENG
University of Ottawa
Ottawa, Ontario, Canada

1969

THE RYERSON PRESS
Toronto

ACADEMIC PRESS
London and New York

ACADEMIC PRESS INC. (LONDON) LTD
BERKELEY SQUARE HOUSE
BERKELEY SQUARE
LONDON, W.1 X.6BA.

U.S. Edition published by
ACADEMIC PRESS INC.
111 FIFTH AVENUE
NEW YORK, NEW YORK 10003

Library of Congress Catalog Card Number: **68-9108**

Filmset by Keyspools Ltd, Golborne, Lancs.
Printed in Great Britain by C. Tinling & Co Ltd,
Liverpool, London and Prescot

To my mother
and my wife

Preface

Threshold Logic has in recent years been intensively studied. There is a growing interest in the subject, and increasing applications of threshold logic elements and networks—owing to the potentiality of a threshold logic element as a general gate in any combinatorial logic, and its suitability as an adaptive element in learning and control systems.

This book is based on the author's notes for a graduate course given at the Department of Electrical Engineering of the University of Ottawa in the past few years.

It is the author's intention to present a clear picture of the basic principles of threshold logic, primarily to electrical engineers, research workers, and graduate students of electrical engineering, who have already had some background in basic switching theory, but are not particularly mathematically oriented. To this end, the author has tried to make the presentation as easily understandable as possible, without sacrifice of mathematical rigour.

Since elementary set theory and Boolean algebra are covered in practically every textbook on general switching circuit theory, they are not included in this book. However, in Chapter 1, the definitions of certain terms in switching theory are given, because they are considered as necessary for the study of threshold logic.

Most of the theorems and their proofs are directly taken from the literature. However, some theorems are slightly modified, reworded, or rearranged, for the sake of clarity and also convenience in sequence of presentation. A few theorems are reproved, even at the expense of length, because the original proofs seem to be difficult to understand, or the reasonings seem to be not very direct.

Chapters 1 through 5 cover the basic properties of threshold functions, and the materials therein are almost all taken from the literature. The line of reasoning and the order of presentation more or less follow those of Dr. Winder in his Ph.D. dissertation and Professor Hu in his book. Chapter 6 is a collection of existing methods of testing and realization. It is not necessarily complete, and some methods are drastically condensed. However, it is believed that these methods are the most representative ones, and that this chapter should give the reader a fairly clear picture of the testing and realization problem.

Chapters 7, 9, and 11, and the major part of Chapter 8, are a collection of the author's own papers on threshold logic, as a result of his research work in the past few years. Chapter 7 is the successive higher ordering method for testing and realization of threshold functions, including that for functions with don't care vertices. The major part of Chapter 8 is the author's own approach to compound synthesis. Chapter 8 also includes a brief description of a method of synthesis of three-input majority gate networks, and an explanation of the basic principle of the realization of symmetric functions with threshold logic elements. Chapter 9, although also the author's own method of analysis of the sensitivity problem, is not too different from other methods in essence, and therefore more or less covers the essential points of the problem. Chapter 10 discusses the most important applications of threshold logic elements. Chapter 11, again the author's own work, presents a method of employing threshold logic elements for probability transformation, which is a very specific but quite interesting application of threshold logic elements.

The author believes that this book has a fairly wide coverage, although it is still far from being complete. For instance, this book does not cover the detailed study of classifications and transformations of functions, orthogonal expansion, multi-threshold logic, threshold functions of order r, etc. However, the author believes that this book gives sufficient materials for a clear and thorough understanding of the basic properties of threshold functions and the testing and realization methods.

No complete bibliography is given, because some of the internal reports on threshold logic are not available to the author. Besides, Professor Hu has already given a fairly complete bibliography in his book. However, important references are given at the end of most chapters.

Acknowledgements

The author wishes to acknowledge that the research work done by him on threshold logic during the past few years was supported in part by the National Research Council of Canada. He wishes to thank Prof. G. S. Glinski, Chairman of the Electrical Engineering Dept. of the University of Ottawa, who initially encouraged him to start work in this field. He appreciates the help of Prof. I. N. Chen and Mr. H. R. Hwa, graduate students in the Electrical Engineering Dept. when the book was written, in reading the manuscript and in correcting minor mistakes. Special thanks are due to Prof. H. C. Torng of Cornell University, who, as a reviewer of the manuscript, gave many valuable suggestions, which have resulted in revision and improvement. Last, but not least, he wishes to express sincere thanks to Dr. S. R. Das, lecturer at the University of Calcutta, India, and presently a postdoctoral fellow with the Electrical Engineering Dept. of the University of Ottawa, who kindly read the proofs twice with utmost care.

September 1968

Ching Lai Sheng
University of Ottawa, Ontario, Canada

Contents

CHAPTER 1

Switching or Boolean Functions

1.1 BASIC DEFINITIONS

In this book we shall study the properties of threshold functions, the necessary and sufficient conditions for a switching function to be a threshold function, the testing and realization methods for threshold functions, and other topics in switching theory that are related to threshold functions.

Since threshold functions are a restricted class of switching functions of binary variables, we start with binary variables and general switching functions. However, those basic concepts such as set theory, switching logic, and Boolean algebra are usually covered in books on general switching theory, and therefore will not be discussed here.

In this chapter we shall give only a minimum amount of review of definitions of terms as considered necessary for the clear understanding of threshold functions, the geometrical interpretation of switching functions in the n-dimensional space, and the clarification of certain notations. For simplicity and convenience we shall use a mixed notation for operations and relations, which is sometimes slightly confusing and therefore needs some explanation.

A *binary variable* x_i is a variable which assumes the value of either element of the two-element set B, i.e.

$$x_i \in B = \{0, 1\}, \text{or } \{-1, 1\}, \text{or } \{F, T\}, \text{or in general, } \{a, b\}.$$

Throughout this book we shall use 0 and 1 for the two elements of B, except in certain special cases, where it is more convenient to use -1 and 1. In this latter case, for instance, the distance from all the vertices to the origin will be the same.

A *literal* is a variable not specified whether it is in the unnegated form (x_i) or in the negated form ($\bar{x}_i$), and is denoted by x_i^*. Loosely we shall call a variable a literal when this variable may be either in the negated form or in the unnegated form.

A *simple product* is a conjunction or product of a subset of the literals $x_1^*, \ldots, x_n^*$, for instance, $x_2\, \bar{x}_4\, x_5$.

A *minterm, fundamental product*, or *n-tuple* is a simple product of all the n literals, for instance, $x_1\, \bar{x}_2 \ldots x_{n-1}\, \bar{x}_n$.

A *switching* or *Boolean function* is a function of n binary variables, such that it assumes a value of either 0 or 1 when each of its n variables assumes a value of 0 or 1.

Note that for n given variables $x_1, x_2, \ldots, x_n$, a function can be independent of $n-1$ variables, and thus reduces to a single variable. For instance, $F = \bar{x}_2$ is such a function. Therefore a single variable, either negated or unnegated, is also considered as a function of n variables.

For n given variables, 0 and 1 are also considered as functions of the n variables. They are independent of all the n variables and are called *constant functions.*

A *Boolean expression, form,* or *representation,* is a particular form in which a Boolean function is expressed. It is built up from the variables and the constants 0 and 1 by a finite number of applications of the Boolean operations addition + and multiplication $\cdot$, and complementation $^{-}$.

A Boolean expression represents a unique switching function of n variables. But a switching function has infinitely many Boolean expressions.

A *normal disjunctive form* (*sum-of-product form*) is a disjunction or sum of simple products.

A *disjunct* is a simple product appearing in a normal disjunctive form.

A *disjunctive canonical form* of a switching function is a normal disjunctive form whose disjuncts are all distinct minterms.

An *implicant* P of a function F is a simple product that can be found in some normal disjunctive forms of F. Obviously P implies F.

An implicant P of a function F is called a *prime implicant* if P implies F, and no other prime implicant P' exists such that P implies P'.

A prime implicant P of a function F is called an *essential prime implicant* if P contains at least one vertex which is not contained in any other prime implicant of F.

An *irredundant normal disjunctive form* is a representation of a switching function as a disjunction of a subset of its prime implicants such that no prime implicant is redundant or may be deleted from the expression.

A *minimum sum-of-product form* is a representation of a switching function as a disjunction of a subset of its prime implicants such that it is minimum according to a certain criterion, say the number of terms in the expression and then the number of literals.

An arbitrary switching function has at least one irredundant normal disjunctive form, and also at least one minimum sum-of-product form. A minimum sum-of-product form is necessarily an irredundant normal disjunctive form, but an irredundant normal disjunctive form is not necessarily a minimum sum-of-product form. If a function has a unique irredundant normal disjunctive form, then the latter is also the unique minimum sum-of-product form of the function.

Any switching function can be expanded along r of its n variables,

$0 \leq r \leq n$, and represented in the form:

$$F = \sum_{i=0}^{2^r-1} G_i F_i,$$

where G_i is a simple product in the given r literals, and i, when written in binary form, designates which of the r variables are to be complemented in G_i by the location of the zeros.

$F_i = F$ when the r variables have assumed the configuration specified by i.

Then $\{F_i\}$ are called the *reduced functions, residues*, or *restrictions* of the given function expanded along the r variables.

EXAMPLE

$$\begin{aligned} F &= x_1x_2x_3+\bar{x}_1x_2x_4+x_1\bar{x}_3\bar{x}_4+x_2\bar{x}_3\bar{x}_4 \\ &= x_1x_2(x_3+\bar{x}_3\bar{x}_4)+x_1\bar{x}_2.\bar{x}_3\bar{x}_4+\bar{x}_1x_2(x_4+\bar{x}_3\bar{x}_4)+\bar{x}_1\bar{x}_2.0 \\ &= x_1x_2\,F_{11}+x_1\bar{x}_2\,F_{10}+\bar{x}_1x_2\,F_{01}+\bar{x}_1\bar{x}_2\,F_{00} \end{aligned}$$

where

$$\begin{aligned} F_{11} &= x_3+\bar{x}_3\bar{x}_4 \\ F_{10} &= \bar{x}_3\bar{x}_4 \\ F_{01} &= x_4+\bar{x}_3\bar{x}_4 \\ F_{00} &= 0 \end{aligned}$$

Since this notation for a reduced function does not always indicate along which r variables the function F is expanded, it is sometimes not clear enough. We shall use a different notation by writing F_{G_i} for F_i. For instance, a function expanded along x_1 and x_2 is written as

$$F = x_1x_2\,F_{x_1x_2}+x_1\bar{x}_2\,F_{x_1\bar{x}_2}+\bar{x}_1x_2\,F_{\bar{x}_1x_2}+\bar{x}_1\bar{x}_2\,F_{\bar{x}_1\bar{x}_2}$$

1.2 GEOMETRICAL INTERPRETATION

Switching functions can be conveniently explained in terms of geometry in the n-dimensional space. As will be seen later, the geometrical interpretation is particularly helpful in the study of threshold functions.

The Cartesian product of n copies of B, $B^n = B \times B \times \ldots \times B$, is called the *n-cube*.

The n-cube B^n is obviously a subset of the continuous Euclidean n-space R^n.

There are 2^n elements in the n-cube, called *vertices*. They are the 2^n different *valuations* of the *ordered n-tuple* $X = (x_1, x_2, \ldots, x_n)$, where x_i, $i = 1, 2, \ldots, n$, is the ith coordinate or component.

A minterm assumes a value of 1 for exactly one vertex or one valuation of the ordered n-tuple. Thus there is a one-to-one correspondance between the minterms and the vertices.

With the commas and parentheses omitted, each vertex is represented by a sequence of 0's and 1's with a total length n, which can be interpreted as a binary number of n digits. A binary number can in turn be represented by its decimal equivalent. Thus, for instance, $10011 = 19$ represents the vertex in the 5-cube with $x_1 = 1$, $x_2 = 0$, $x_3 = 0$, $x_4 = 1$, and $x_5 = 1$.

In the geometrical sense, a switching function of n variables, $F(x_1, \ldots, x_n)$, is defined by assigning either 0 or 1 to the 2^n vertices of B^n, i.e. a mapping from the n-cube B^n into B, or

$$F : B^n \to B.$$

Throughout this book we shall denote switching functions by F and ordinary functions by f.

For a function F, the set of vertices in the n-cube that is mapped into 1 is called the *on-set* of F, and the set of vertices in the n-cube that is mapped into 0 is called the *off-set* of F. Since B consists of only two elements, F is uniquely determined by specifying its on-set alone.

For simplicity, we shall not distinguish between a minterm and a vertex, nor shall we distinguish between a switching function and its on-set. Thus we shall denote a switching function either by a Boolean expression or by its on-set, without making the rigorous distinction. For instance,

$$\begin{aligned} F(x_1, x_2, x_3) &= x_1\bar{x}_2 + x_1x_3 \\ &= x_1\bar{x}_2\bar{x}_3 + x_1\bar{x}_2x_3 + x_1x_2x_3 \\ &= \{100, 101, 111\} \\ &= \{4, 5, 7\} \end{aligned}$$

In an n-cube, the subset of vertices corresponding to a Boolean product of n-r literals, $0 \le r \le n$, where each literal is present as a factor exactly once, either complemented or not, is called an *r-cube*. Obviously an r-cube consists of 2^r vertices.

r is called the *dimension* of the r-cube.

Since as a subset of vertices, an r-cube is represented by a sequence of 1's and 0's, the coordinate index i of x_i^* present in the Boolean product is not shown. So we shall use a don't care symbol * in those places of x_j^*, where x_j^* is not present in the product. For instance, in a 5-cube, the 2-cube corresponding to $x_1\bar{x}_3x_4$ is represented as 1*01*. Since we do not distinguish a minterm from a vertex, we shall also not distinguish a simple product from an r-cube. Thus the r-cube 1*01* is also denoted by $x_1\bar{x}_3x_4$.

Two r-cubes A and D in B^n are said to be *consecutive* if they differ in exactly one coordinate.

A 0-cube consists of a single vertex.

The 1-cubes in B^n are called the *edges* of B^n.

Any two consecutive vertices of B^n form an edge of B^n.

The 2-cubes in B^n are called the *squares* of B^n.

If we replace one of the two *'s in the coordinates of a square by 1 or 0, respectively, we obtain two consecutive edges of B^n called a pair of *opposite sides* of the square.

For $r \geq 3$, if we replace one of the r*'s in the coordinates of an r-cube by the integers 1 or 0, respectively, we obtain two consecutive $(r-1)$-cubes called a pair of *opposite faces* of the r-cube.

1.3 MIXED NOTATION FOR OPERATIONS AND RELATIONS

It can be shown that the set ϕ of all subsets of the n-cube B^n is a Boolean algebra, and that it is isomorphic with the Boolean algebra Φ of all switching functions of n variables. For simplicity and convenience, we shall not distinguish between them. Also, to conform with the most common usage in literature, we shall adopt a mixed notation for Boolean and set-theoretic operations and relations. The symbols used are as follows:

- $+$ denotes Boolean addition and also ordinary algebraic addition.
- $\cdot$ denotes Boolean multiplication and also ordinary algebraic multiplication. It is sometimes omitted when there is no ambiguity.
- $\bar{}$ denotes complementation or negation, e.g. $\bar{x}_1$, $\bar{F}$.
- $=$ denotes equivalence.
- $\subset$ denotes upper bound property for sets and also for switching functions. It reads "is contained in", and is used in place of the symbol $\Rightarrow$ for "implies".
- $\supset$ denotes lower bound property for sets and also for switching functions. It reads "contains", and is used in place of the symbol $\Leftarrow$ for "is implied by".
- $\leq$ denotes algebraic upper bound property.
- $\geq$ denotes algebraic lower bound property.

CHAPTER 2

Definition of Threshold Function and Basic Concepts of Linear Separation

2.1 DEFINITIONS

In this chapter we shall give the definitions of threshold function and those related terms such as linear separability, realization, threshold logic element, etc. We shall explain linear separability in terms of convex sets and convex hulls, and shall explain the derivation of the upper bound of number of threshold functions in terms of dual spaces. Also we shall discuss the geometrical concept of region or connectedness, as a weak but easy-to-visualize necessary condition for linear separability.

A Boolean function $F(X)$ of n binary variables X is said to be a *threshold function* if the following conditions are satisfied:

$$F(X) = 1, \text{ if } f_A(X) = \sum_{i=1}^{n} a_i x_i \geq T \tag{2.1–1}$$

$$F(X) = 0, \text{ if } f_A(X) = \sum_{i=1}^{n} a_i x_i < T \tag{2.1–2}$$

where

x_i = a binary variable assuming a value of either 0 or 1, for $i = 1, 2, \ldots, n$.
$X = (x_1, x_2, \ldots, x_n)$
= an ordered n-tuple or vector on n binary variables.
a_i = a real coefficient called the *weight* of x_i, for $i = 1, 2, \ldots, n$.
$A = (a_1, a_2, \ldots, a_n)$
= an ordered n-tuple or vector of n coefficients for the n variables in the algebraic function, or the *weight vector*.
T = a constant called the *threshold value*.
$F(X)$ = a Boolean function of X.
$f_A(X)$ = an algebraic function of X.

Threshold function is also called *linearly separable function*, *1-realizable function*, *linear-input function*, *majority decision function*, etc.

The term "linearly separable function" means that, in geometrical interpretation, in the n-dimensional space the set of vertices represented by F can be separated from the set of vertices represented by the complementary function $\bar{F}$ by a *hyperplane*. The term "1-realizable function" means that the function can be realized by a *single* threshold logic element. The term "linear-input function" means that the algebraic equation representing the separating surface contains only first-order terms and therefore is *linear*. The term "majority decision function" is slightly misleading. Strictly it should mean the function which can be realized by a majority gate, a gate whose output is 1 if and only if a majority of the inputs are 1, for instance, a 2 out of 3 majority gate. However, in the literature, the term majority decision function has been used for a general threshold function.

A threshold logic element is a physical element whose inputs consist of n binary variables $\{x_1, x_2, \ldots, x_n\}$ and whose output F is a threshold function of the n variables.

In other words, a threshold logic element is a physical realization of a threshold function.

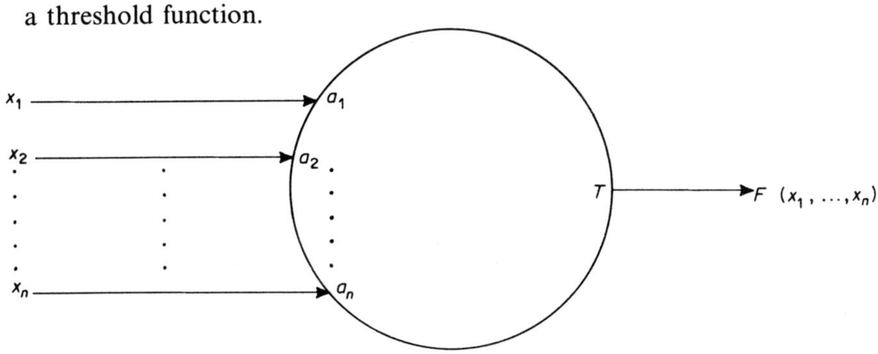

FIG. 1. Symbol for a threshold logic element.

There are several symbols for a threshold logic element. Here we shall adopt the most popular one, as shown in Fig. 1. The weights of the input variables, $\{a_i\}$, and the threshold value T are indicated inside the circle.

Threshold logic element is also called *threshold logic unit*, *linearly separable device*, *threshold device*, *linear-input device*, etc.

1-*realizability* (*linear separability*) is the capability of a Boolean function to be realized by a single threshold logic element. Thus a threshold function is 1-realizable.

Considering X and A as vectors, we can see that $f_A(X)$ is the inner product of X and A. If X and A are considered as $1 \times n$ matrices, then $f_A(X)$ is the product of X and A^t or the product of A and X^t, where X^t and A^t are the transposes of X and A, respectively.

The n-cube consists of 2^n vertices. If a function F of n variables has m vertices, then the complementary function $\bar{F}$ has $m' = 2^n - m$ vertices.

Let

$X_j, j = 1, 2, \ldots, m$, be a vertex of F,

$X_k, k = m+1, m+2, \ldots, m+m' = 2^n$, be a vertex of $\bar{F}$.

Then a threshold function can be defined as follows:

A Boolean function F of n variables is a threshold function if there exists an $n \times 1$ matrix A^t of real numbers and a real threshold value T such that

$$X_j A^t \geq T, \quad j = 1, 2, \ldots, m, \tag{2.1–3}$$

$$X_k A^t < T, \quad k = m+1, m+2, \ldots, m+m' = 2^n. \tag{2.1–4}$$

It is easily seen that $f_A(X) = XA^t$ is a mapping from the n-cube into the one-dimensional line of real numbers. Let

$X_j A^t_{(min)}$ be the minimum value of all $X_j A^t$'s,

$X_k A^t_{(max)}$ be the maximum value of all $X_k A^t$'s,

T_u be the upper boundary of T,

T_l be the lower boundary of T.

Then from the above definition, we have

$T_u = X_j A^t_{(min)}$,

$T_l = X_k A^t_{(max)}$,

$T_l < T \leq T_u$.

Note that the set of possible values of T is closed at the upper boundary, but is open at the lower boundary.

$g = T_u - T_l$ is called the *gap length.*

EXAMPLE

$F(x_1, x_2, x_3) = x_1 x_2 + x_1 \bar{x}_3$

j	X_j	k	X_k
1	100	4	000
2	110	5	001
3	111	6	010
		7	011
		8	101

It is known that F is a threshold function. One possible set of weights and threshold value is

$$a_1 = 2, a_2 = 1, a_3 = -1, T = 2$$

The mapping $f_A(X)$ is shown in Fig. 2.

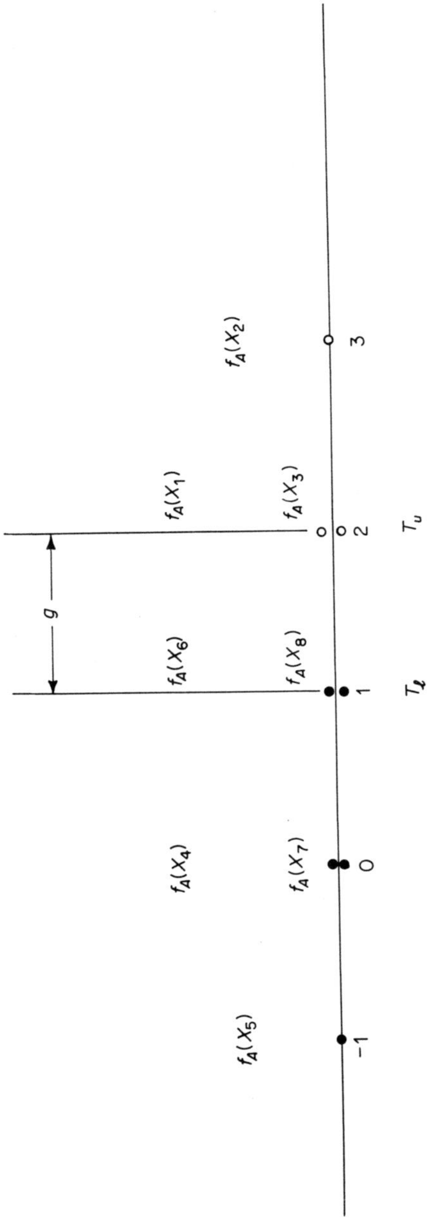

FIG. 2. Mapping of $F = x_1x_2 + x_1\bar{x}_3$ with $A = (2, 1, -1)$ and $T = 2$.

The vector A and the value T completely determine the threshold function. We shall call the set of ordered numbers consisting of a_1, a_2, ...,

a_n, and T a 1-*realization*, or simple *realization*, of F, and represent it as $R[F]$. Thus

$$R[F] = (a_1, a_2, \ldots, a_n; T).$$

There are numerous 1-realizations for any threshold function. For instance, for an *AND* gate of three variables x_1, x_2, and x_3, we may have $R_1[F] = (1, 1, 1; 3)$, $R_2[F] = (1, 1, 2; 4)$, $R_3[F] = (2, 2, 2; 6)$, etc.

The equation

$$XA^t = T$$

represents a hyperplane in the n-dimensional Euclidean space, called a *separating hyperplane*, or *separating plane*. This is why a threshold function is also called a linearly separable function.

The realization $R[F] = (a_1, a_2, \ldots, a_n; T)$ is also called a *separating system*.

When $T_l < T < T_u$, the separating system is called a *strict separating system*, or *strict realization*.

In the following, we shall not distinguish between a separating system and a strict separating system, unless it is necessary.

2.2 INTEGRAL REALIZATION

Let $R[F] = (a_1, a_2, \ldots, a_n; T)$ be a realization of a threshold function F. Since the rational numbers are everywhere dense in the space R of all real numbers, we can choose $a_1, a_2, \ldots, a_n$, and T to be rational numbers. Now let m be the lowest common multiple of the denominators of the $n+1$ rational numbers. Set

$$a_i' = m\,a_i, \quad i = 1, 2, \ldots, n,$$
$$T' = m\,T.$$

Then $a_1', a_2', \ldots, a_n'$, and T' are integers. Thus we can conclude that for every threshold function F, there exists an *integral realization* $R[F] = (a_1', a_2', \ldots, a_n'; T')$, where the weights and the threshold value are all integers.

In the study of threshold functions, it is much more convenient to consider integral realization than fractional realization. Besides, as will be discussed later, it is sufficiently general to consider integral realization only, because usually the *minimal-integral realization* is also the *minimal realization*, except in certain very special cases.

2.3 CONVEX SETS

The necessary and sufficient condition for a switching function F of n variables to be a threshold function has been studied extensively and has been expressed in various forms.[1, 3, 6, 8, 12, 14, 15, 17, 19] However, they all seem to originate from one basic concept, *viz.* the geometric concept of convex sets and convex hulls.[1, 4, 6, 8, 9] Apart from serving as a necessary and sufficient condition, the geometric concept is also helpful in visualization and in the study of the number of threshold functions. Here we shall study the basic properties of convex sets and duality in n-dimensional space.

DEFINITION. A set C is said to be *convex* if, whenever X_1 and X_2 are points of C, the entire line segment between X_1 and X_2 also belongs to C.

An arbitrary point X_i on the segment between X_1 and X_2 can be expressed as

$$X_i = c_i X_1 + (1 - c_i) X_2, \quad 0 \leq c_i \leq 1. \tag{2.3–1}$$

DEFINITION. If A_j is a fixed point in the n-dimensional space and T is a number, then the truth sets of inequalities of the form $XA_j^t > T$ or $XA_j^t < T$ are called *open half-spaces*, while the truth sets of the inequalities $XA_j^t \geq T$ or $XA_j^t \leq T$ are called *closed half-spaces* in this space. The truth set of the equation $XA_j^t = T$ is called a *hyperplane*.

The following theorems are well known.

THEOREM 2.1 The intersection of two or more convex sets is a convex set.

THEOREM 2.2 A half-space, either closed or open, is convex.

DEFINITION. The intersection of a finite number of closed half-spaces passing through the origin is called a *convex cone*.

DEFINITION. The *convex hull* of an arbitrary set of points F in the Euclidean space R^n, $H(F)$, is the smallest convex set which contains F.

A convex hull consists of all points of R^n of the form

$$X = c_1 X_1 + c_2 X_2 + \ldots + c_p X_p, \tag{2.3–2}$$

where $c_1, \ldots, c_p$ are non-negative real numbers satisfying

$$\sum_{i=1}^{p} c_i = 1.$$

There is no precise definition of *duality* in Euclidean space and therefore there is a choice of dual spaces available.

A *dual space* to an n-dimensional Euclidean space R^n is a space in which the hyperplanes or half-spaces of R^n are represented by points or half-lines.

We shall consider the space of weights as dual of the space of variables, and the *variable space* as a dual of the *weight space*.

Let R^n be the n-dimensional Euclidean space of the variables and $(R^n)^*$ be the dual n-dimensional space of the corresponding weights. If F

is any subset of R^n, then the dual of F is that subset W of $(R^n)^*$, such that

$$XA^t \geq T$$

where $X \in F \subset R^n$, $A \in W \subset (R^n)^*$, and $T =$ a constant.

If X is a single point $X_1 = (x_{11}, x_{12}, \cdots, x_{1n})$, then W is a closed half-space in the weight space $(R^n)^*$.

If inequalities $XA^t \geq T$ have to be satisfied simultaneously for a certain set of points $\{X_j\}$, then the dual of $\{X_j\}$ is the intersection of the half-spaces which are dual to the X_j's.

When $T = 0$, the boundary of a half-space is a hyperplane represented by $XA^t = 0$. (2.3–3)

In general the number of these hyperplanes is equal to the number of possible points. However, in the case when the point is a zero vector, there is no hyperplane because any vector in weight space will satisfy Equation (2.3–1). Thus we can say that the whole space is divided into a certain number of convex sets by $(2^n - 1)$ hyperplanes. Each of these hyperplanes has as its normal a variable vector and passes through the origin. Therefore the weight space is divided into a certain number of cones. When two different weight vectors lie on the same side of each one of the $2^n - 1$ hyperplanes, they belong to a certain cone and will realize the same function.

The concept of weight space and convex cone is very helpful in the determination of the number of threshold functions, as will be studied in the next section.

There is a theorem in the theory of convex sets, which states that two convex hulls in R^n are strictly separated by a hyperplane iff (if and only if) they are disjoint. Then we can easily prove the following theorem.

THEOREM 2.3–1. The switching function F is a threshold function iff

$$H(F) \bigcap H(\bar{F}) = \emptyset,$$

where $\emptyset$ is the empty set.

PROOF: *Necessity.* Assume F to be a threshold function. Then there exists a hyperplane which strictly separates F from $\bar{F}$. Since the half-spaces are convex, they contain the hulls. Therefore the hyperplane also strictly separates $H(F)$ from $H(\bar{F})$. By the above theorem in convex sets theory, $H(F)$ and $H(\bar{F})$ are disjoint.

Sufficiency. Assume that $H(F)$ and $H(\bar{F})$ are disjoint. By the above theorem, there exists a hyperplane which strictly separates $H(F)$ from $H(\bar{F})$. Since F and $\bar{F}$ are contained in $H(F)$ and $H(\bar{F})$, respectively, the hyperplane strictly separates F from $\bar{F}$. Therefore F is a threshold function.

Q.E.D.

THEOREM 2.3–1 is very basic and important. It leads directly to Chow's theorems and Elgot's assumability, as will be studied later.

2.4 NUMBER OF THRESHOLD FUNCTIONS

Several upper and lower bounds of the number of threshold functions have been derived.[13, 16, 18, 20, 21] Here we shall study the number of threshold functions from the point of view of weight space,[7] and shall give an upper bound only.

For the study of the number of threshold functions, an n-input n-weight threshold logic element is considered as an $(n+1)$-input $(n+1)$-weight element, with the $(n+1)$ th input $x_{n+1} = 1$ and the $(n+1)$ th weight $a_{n+1} = -T$.
Thus we have

$$X = (x_1, x_2, \ldots, x_n, 1),$$

$$A = (a_1, a_2, \ldots, a_n, -T).$$

In this case the set of points in input space also consists of 2^n different input combinations.

The $(n+1)$-dimensional weight space $(R^{n+1})^*$ is divided into a number of cones, by 2^n hyperplanes, each of which corresponds to one of the 2^n input vectors. Each of these cones is a region which contains all the weight vectors of a threshold function.

The number of cones obtained when the $(n+1)$-dimensional space is partitioned by 2^n hyperplanes is, in general, larger than that obtained when the n-dimensional space is partitioned by the same number of hyperplanes. Therefore the number of functions which can be realized by $(n+1)$-weights element is larger than that in the case of n-weights element.

The general problem of how many threshold functions there are for n variables remains unsolved at present. The calculation of an upper bound for the number of threshold functions, $R(n)$, however, can be done by considering the problem of the maximum number of cones into which any number of hyperplanes passing through the origin may divide a space of any dimensions.

Let $C_{m,n}$ be the maximum number of cones into which m hyperplanes passing through the origin may divide a space of n dimensions. It is obvious that $C_{m,1} = 2$ for each $m > 0$, $C_{m,2} = 2m$, $C_{1,n} = 2$ for any $n > 0$.

The general formula can be derived by the following argument. Suppose that formula has been established for $m-1$ hyperplanes in the n-dimensional space. The mth hyperplane will be divided by $m-1$ hyperplanes (along at most $m-1$ hyperplanes) into $C_{m-1,n-1}$ pieces. Each of these hyperplanar pieces divides the region it belongs to into two new regions, i.e. out of the original $C_{m-1,n}$ regions at most $C_{m-1,n-1}$ regions are doubled. Therefore we have

$$C_{m,n} = C_{m-1,n-1} + C_{m-1,n}. \tag{2.4–1}$$

For $m \leq n$, Equation (2.4–1) can be expanded as follows:

$$\begin{aligned} C_{m,n} &= C_{m-1,n-1} + C_{m-1,n} \\ &= C_{m-2,n-2} + 2C_{m-2,n-1} + C_{m-2,n} \\ &= \dots \\ &= \binom{m-1}{0} C_{1,n-m+1} + \binom{m-1}{1} C_{1,n-m+2} + \dots \\ &+ \binom{m-1}{m-1} C_{1,n}, \end{aligned} \tag{2.4–2}$$

where $\binom{j}{i} = \dfrac{j!}{i!\,(j-i)!}$.

Since $C_{1,n} = 2$ for any integer $n > 0$, Equation (2.4–2) becomes

$$\begin{aligned} C_{m,n} &= 2\binom{m-1}{0} + 2\binom{m-1}{1} + \dots + 2\binom{m-1}{m-1} \\ &= 2 \times 2^{m-1} \\ &= 2^m. \end{aligned} \tag{2.4–3}$$

For $m > n$, Equation (2.4–1) can be expanded as follows:

$$\begin{aligned} C_{m,n} &= \binom{m-1}{0} C_{1,n-m+1} + \dots + \binom{m-1}{m-n} C_{1,1} \\ &+ \binom{m-1}{m-n+1} C_{1,2} + \dots + \binom{m-1}{m-1} C_{1,n} \end{aligned} \tag{2.4–4}$$

Since $C_{1,n} = 0$ for any integer $n \leq 0$ and since $\binom{j}{i} = \binom{j}{j-i}$, Equation (2.4–4) becomes

$$\begin{aligned} C_{m,n} &= 2\binom{m-1}{m-n} + 2\binom{m-1}{m-n+1} + \dots + 2\binom{m-1}{m-1} \\ &= 2\binom{m-1}{n-1} + 2\binom{m-1}{n-2} + \dots + 2\binom{m-1}{0} \\ &= 2\sum_{k=0}^{n-1} \binom{m-1}{k}. \end{aligned} \tag{2.4–4′}$$

Thus the number of threshold functions realized by threshold logic elements of n inputs and n weights is

$$R(n) \leq 2\sum_{k=0}^{n-1} \binom{2^n-1}{k}. \tag{2.4–5}$$

If there are $n+1$ weights and n inputs,

$$R(n+1) \leq 2\sum_{k=0}^{n} \binom{2^n-1}{k}. \tag{2.4–6}$$

Another formula for both the lower and the upper bound of the number of threshold functions, as derived by Winder[19], is

$$2^{0\cdot 33n^2} < R(n)\, 2^{n^2} \leqslant 2^{2^n}, (n \gg 1). \tag{2.4–7}$$

The ratio of $R(n)$ to the total number of functions of n variables decreases rapidly as n increases.

Table I gives a comparison of the number of threshold functions and the total number of functions for n up to six.[19]

TABLE I

Number of threshold functions

Number of variables, n	Number of switching functions, 2^{2^n}	Number of threshold functions, $R(n)$
0	2	2
1	4	4
2	16	14
3	256	104
4	65,536	1882
5	$\sim 4\cdot 3 \times 10^9$	94,572
6	$\sim 1\cdot 8 \times 10^{19}$	15,028,134

2.5 REGION CONCEPTS

Kaszerman[10, 11] has studied threshold functions from the geometric point of view using region concepts. Here we shall study region concepts briefly.

DEFINITION. Two vertices of the n-cube B^n are said to be *adjacent* if they differ in exactly one coordinate or have a Hamming distance of one between them.

It is obvious that any two adjacent vertices of B^n are precisely consecutive 0-cubes in B^n.

DEFINITION. A set K in B^n is said to be a *region* or to be *connected*[9] iff for any two vertices X and Y in K there exists a finite sequence of vertices $V_0, V_1, \ldots, V_q$ in K such that $V_0 = X$, $V_q = Y$, and for each $i = 1, 2, \ldots, q$, the pair of vertices V_{i-1}, V_i are adjacent.

The property of a set being a region is called *connectedness.*

DEFINITION. A proper subset S of a region R is said to be a *subregion* of R iff S itself is also a region.

DEFINITION. An *interior vertex* of a region is a vertex such that it is a member of the region and that all its adjacent vertices are also members of the region.

DEFINITION. A *boundary vertex* is a vertex such that it is a member of a region and that at least one of its adjacent vertices is not a member of the region.

DEFINITION. An *exterior vertex* of a region is a vertex which is not a member of the region.

As will be defined in Chapter 4, a switching function F of n variables is a *unate function* if there exists an expression for F such that each variable appears in either unnegated or negated form but not both.

THEOREM 2.5–1. The vertices of a unate function form a region.

PROOF: If a unate function is expressed in an irredundant normal disjunctive form, each variable appears only either unnegated or negated, and each disjunct is a prime implicant. Each prime implicant represents an r-cube, which is obviously a region. It can be easily shown that all the prime implicants have at least one vertex in common. Therefore the vertices of a unate function is a region. Q.E.D.

COROLLARY 2.5–1–1. The vertices of a threshold function form a region.

PROOF: It will be proved in Chapter 4 that a threshold function is necessarily a unate function. Therefore the vertices form a region. Q.E.D.

For convenience the boundary vertices of a function F will be called the *true boundary vertices*, and the boundary vertices of $\bar{F}$ will be called the *false boundary vertices*.

THEOREM 2.5–2. A necessary and sufficient condition that a Boolean function F be a threshold function is that the true boundary vertices be separable from the false boundary vertices by a hyperplane.

PROOF: *Necessity.* Necessity follows directly from the definition since the boundary vertices were defined as part of F or $\bar{F}$.

Sufficiency. Assume that there exists a hyperplane which separates the true boundary vertices from the false boundary vertices, i.e. there exists an A such that

$$XA^t \geq T \text{ at a true boundary vertex.}$$
$$XA^t < T \text{ at a false boundary vertex.}$$

Without loss of generality, we assume that the weights or components of A are all positive. Then ignoring the trivial cases that $F = 0$ or 1,

$$(1, 1, \ldots, 1) \text{ is a true vertex,}$$
$$(0, 0, \ldots, 0) \text{ is a false vertex.}$$

Next, an interior true vertex is examined. Assume k of its components are 0. Renumber the variables so that it is the first k variables which are 0, i.e.

$$X_{interior} = (0, 0, \ldots, k^{\text{th}} \; 0, 1, \ldots, 1).$$

Then

$$X_{interior}A^t = a_{k+1} + \ldots + a_n. \quad (2.5\text{–}1)$$

Now consider a sequence of vertices such that the first is the $X_{interior}$ and each succeeding vertex has an additional component changed to 0. At least one of these must be a boundary true vertex because the sequence starts with a true vertex and reaches the false vertex (0, 0, ..., 0) in unit steps. Assume that the boundary true vertex is the one which has 0 for the first $p(p > k)$ components. Call this vertex $X_{boundary}$. Then

$$X_{boundary}A^t = a_{p+1} + \ldots + a_n. \quad (2.5\text{–}2)$$

Comparing (2.5–1) and (2.5–2), and by the fact that all a_i are positive and $p > k$, we have

$$X_{interior}A^t > X_{boundary}A^t \geq T.$$

Thus, an interior true vertex is on the same side of the hyperplane as the boundary true vertex.

By a similar argument, it can be shown that the interior false vertices are on the same side of the hyperplane as the boundary false vertex.

Therefore a hyperplane which separates the boundary vertices will correctly separate all the F vertices from all the $\bar{F}$ vertices. Q.E.D.

COROLLARY 2.5.2–1. A reduced set of inequalities which defines a threshold function, and therefore is equivalent to the complete set of inequalities, can be formed by using only the boundary true and false vertices of a Boolean function.

Connectedness is a rather weak condition, which is far from being sufficient for a threshold function. However, the region concept does give a clear geometrical interpretation, and has been utilized by Kaszerman for the synthesis of two-threshold-logic-element networks.

REFERENCES

1. Chow, C. K. (1961). Boolean functions realizable with single threshold devices, *Proc. Inst. Radio Engrs* **49,** 270–271.
2. Coates, C. L., and Lewis, P. M. (1961). Linearly separable switching functions, *J. Franklin Inst.* **272,** 360–410.
3. Dertouzos, M. L. (1964). An approach to single-threshold-element synthesis, *I.R.E. Trans. electronic Comput.* EC **13,** 519–528.
4. Eggleton, H. G. (1958). "Convexity", Cambridge University Press, Cambridge, England.
5. Elgot, C. C. (1961). Truth functions realizable by single threshold organs, in "Switching Circuit Theory and Logical Design", *A.I.E.E. Special Publication* **S134,** 225–245.

6. Gabelman, I. J. (1962). A note on the realization of Boolean function using a single threshold element, *Proc. Inst. Radio Engrs* **50,** 225–226.
7. Glinski, G. S., and Zawicki, I. J. (1964). Properties of a single 'linear' threshold logic unit, *Technical Report* No. 64–14, Department of Electrical Engineering, University of Ottawa.
8. Highleyman, H. (1961). A note on linear separation, *I.R.E. Trans. electronic Comput.* EC **10,** 777–778.
9. Hu, S. T. (1965). "Threshold Logic", University of California Press, U.S.A.
10. Kaszerman, P. (1963). A geometric test-synthesis procedure for a threshold device, *Inf. Control* **6,** 381–398.
11. Kaszerman, P. (1965). A 'region' concept and its application to threshold logic, *Inf. Control* **8,** 531–551.
12. Muroga, S. (1961). Functional forms of majority decision functions and a necessary and sufficient condition for their realizability, in "Switching Circuit Theory and Logical Design", *A.I.E.E. Special Publication* S**134,** 39–46.
13. Muroga, S. (1965). Lower bounds of the number of threshold functions and a maximum weight, *I.E.E.E. Trans. electronic Comput.* EC **14,** 136–148.
14. Muroga, S., Toda, I., and Takasu, S. (1961). Theory of majority decision elements, *J. Franklin Inst.* **271,** 376–418.
15. Paull, M. C., and McCluskey, E. J. Jr. (1960). Boolean functions realizable with single threshold devices, *Proc. Inst. Radio Engrs* **48,** 1335–1337.
16. Smith, D. R. (1966). Bounds on the number of threshold functions, *I.E.E.E. Trans. electronic Comput.* EC **15,** 368–369.
17. Winder, R. O. (1961). Single stage threshold logic, in "Switching Circuit Theory and Logical Design", *A.I.E.E. Special Publication* S**134,** 321–332.
18. Winder, R. O. (1961). More about threshold logic, in "Switching Circuit Theory and Logical Design", *A.I.E.E. Special Publication* S**134,** 55–64.
19. Winder, R. O. (1962). "Threshold Logic", Doctoral dissertation, Mathematics Department, Princeton University.
20. Winder, R. O. (1963). Bounds on threshold gate realizability, *I.E.E.E. Trans. electronic Comput.* EC**12,** 561–564.
21. Yajima, S., and Ibaraki, T. (1965). A lower bound of the number of threshold functions, *I.E.E.E. Trans. electronic. Comput.* EC **14**, 926–929.

CHAPTER 3

Characterization of Threshold Functions

3.1 r-CUBES IN B^n

In this chapter we shall study the characterization of threshold functions by a set of parameters. As a preparation, we shall study certain properties of r-cubes in B^n and reduced functions further.

The ith coordinate of a cube C in B^n is said to be a *free coordinate* of C iff it is a *; otherwise it is called a *fixed coordinate.*

The dimension of C, r, is the number of free coordinates, and the *codimension* of C, $n-r$, is the number of fixed coordinates.

Several cubes are said to be *isomerous* iff they have exactly the same free coordinates.

For instance, for $n = 5$, $x_1\bar{x}_3x_4 = 1{*}01{*}$ and $\bar{x}_1x_3x_4 = 0{*}11{*}$ are isomerous.

It is obvious that for every set of r free coordinates in B^n, there are 2^{n-r} distinct and disjoint isomerous r-cubes whose union is B^n.

Two r-cubes in B^n are said to be *opposite* iff they are isomerous and have complementary fixed coordinates.

A pair of vertices, $X = x_1, \ldots, x_n$ and $Y = y_1, \ldots, y_n$ in an r-cube C is said to be a *diagonal* of C iff for every free coordinate i, $x_i = \bar{y}_i$.

For instance, for $n = 5$, in the 2-cube $x_1\bar{x}_2$, the two vertices $x_1\bar{x}_2x_3\bar{x}_4x_5$ and $x_1\bar{x}_2\bar{x}_3x_4\bar{x}_5$ are a diagonal.

3.2 REDUCED FUNCTIONS

As is well known, a Boolean function of n variables, F, can be expanded or decomposed along any variable x_i as

$$F = x_iF_{x_i} + \bar{x}_iF_{\bar{x}_i}$$

where

$$F_{x_i} = F(x_1, \ldots, x_{i-1}, 1, x_{i+1}, \ldots, x_n)$$

$$F_{\bar{x}_i} = F(x_1, \ldots, x_{i-1}, 0, x_{i+1}, \ldots, x_n)$$

Now F_{x_i} and $F_{\bar{x}_i}$ can in turn be expanded along any other variable x_j.

Thus a function can be expanded along any subset of variables. For a set of two variables, $\{x_i, x_j\}$,

$$F = x_i x_j F_{x_i x_j} + x_i \bar{x}_j F_{x_i \bar{x}_j} + \bar{x}_i x_j F_{\bar{x}_i x_j} + \bar{x}_i \bar{x}_j F_{\bar{x}_i \bar{x}_j}$$

It is seen that in the n-cube x_i and $\bar{x}_i$ represent two opposite $(n-1)$-cubes. Similarly, in the n-cube, $x_i x_j$, $x_i \bar{x}_j$, $\bar{x}_i x_j$, and $\bar{x}_i \bar{x}_j$ represent four isomerous $(n-2)$-cubes. $x_i F_{x_i}$ consists of the subset of vertices of F in the $(n-1)$-cube x_i; $\bar{x}_i F_{\bar{x}_i}$ consists of the subset of vertices of F in the $(n-1)$-cube $\bar{x}_i$. Similarly $x_i x_j F_{x_i x_j}$ consists of the subset of vertices of F in the $(n-2)$-cube $x_i x_j$, etc.

3.3 CHARACTERIZING PARAMETERS

Chow[1] found a set of $n+1$ parameters to uniquely characterize the realization of n-variable threshold functions, as are called the *Chow parameters.*

As mentioned above, F denotes a Boolean function and also the on-set of F in the n-cube B^n, and $\bar{F}$ denotes the complementary function of F and also the off-set of F in the n-cube.

Let m be the number of vertices of F or true vertices, and $m' = 2^n - m$ be the number of vertices of $\bar{F}$ or false vertices. Consider the vertices as n-dimensional vectors. Let

$X_j = (x_{j1}, x_{j2}, \ldots, x_{jn}), j = 1, 2, \ldots, m$, be a true vertex.

$X_k = (x_{k1}, x_{k2}, \ldots, x_{kn})$, $k = m+1,\ m+2,\ \ldots,\ m+m' = 2^n$, be a false vertex.

$C = \sum_{j=1}^{m} X_j = (c_1,\ c_2,\ \ldots,\ c_n)$ be the vector sum of all true vertices.

Then

$$c_i = \sum_{j=1}^{m} x_{ji}.$$

Since x_{ji} is either 0 or 1, c_i is a nonnegative integer. It is obvious that when x_{ji} is 1, X_j is a vertex in the $(n-1)$-cube x_i, and when $x_{ji} = 0$, X_j is a vertex in the $(n-1)$-cube $\bar{x}_i$. Thus c_i is equal to the number of true vertices in the $(n-1)$-cube x_i, or the total number of vertices in $x_i F_{x_i}$. Also $m - c_i$ is equal to the number of true vertices in the $(n-1)$-cube $\bar{x}_i$, or the total number of vertices in $\bar{x}_i F_{\bar{x}_i}$.

C/m is called the *centre of gravity* of F. It is obviously a point in the n-dimensional Euclidean space, but not a vertex of the n-cube. Besides, it belongs to the hull $H(F)$.

Two switching functions F and G are said to be *equipollent* iff they have the same parameters m, c_1, …, and c_n.

Two functions or sets F and G are said to be *similar*[4] iff

$$\frac{c_i(F)}{m(F)} = \frac{c_i(G)}{m(G)} \text{ for every } i = 1, 2, \ldots, n.$$

In other words, F and G are similar iff their centres of gravity are the same.

It is obvious that two functions equipollent to each other are necessarily similar, but not vice versa. It is also obvious that the functions corresponding to two pairs of diagonals of the same r-cube are also equipollent and similar.

3.4 THEOREMS RELATED TO THE CHARACTERIZING PARAMETERS[1,4,5,6]

THEOREM 3.4.1. Let F be a threshold function of n variables and A and B be two similar functions. If F contains A, then F must contain at least one vertex of B.

PROOF: Suppose that F does not contain any vertex of B. Then $\bar{F}$ contains B. By Theorem 2.3.1, $H(F)$ and $H(\bar{F})$ are disjoint. Therefore $H(A)$ and $H(B)$ are disjoint too, and A and B cannot have the same centre of gravity. This contradicts the fact that A and B are similar. Therefore F must contain at least one vertex of B. Q.E.D.

COROLLARY 3.4.1–1. If a threshold function F of n variables contains a diagonal of an r-cube C, then it must contain at least one vertex of every diagonal of C.

THEOREM 3.4.2. Let two switching functions of n variables, F and G, be equipollent. If either F or G is a threshold function, then $F = G$.

PROOF: Suppose $F \neq G$. Since $m(F) = m(G)$, $F\bar{G}$ and $\bar{F}G$ are both non-empty and consist of equal number of vertices. Furthermore, since $C(F) = C(G)$, $F\bar{G}$ and $\bar{F}G$ are similar and their centres of gravity are equal.

Since F is a threshold function, by Theorem 2.3.1, $H(F)$ and $H(\bar{F})$ are disjoint. Therefore $H(F\bar{G})$ and $H(\bar{F}G)$ are disjoint too, and $F\bar{G}$ and $\bar{F}G$ cannot have the same centre of gravity, a contradiction. Hence $F = G$. Q.E.D.

COROLLARY 3.4.2–1. Let two switching functions F and G be equipollent. If either F or G is not a threshold function, the other is also not a threshold function.

COROLLARY 3.4.2–2. If two distinct functions F and G are equipollent, then both F and G are not threshold functions.

COROLLARY 3.4.2–3. A threshold function F is completely determined by its parameters $m, c_1, c_2, \ldots, c_n$.

The parameters $m, c_1, c_2, \ldots, c_n$ completely characterize a threshold function. Because of the uniqueness, they can be employed to label known threshold functions. $m, c_1, \ldots, c_n$ are called the Chow parameters.

3.5 PRIME FUNCTIONS

If for a switching function F of n variables, $m \geq 2^{n-1}$, then for the complementary function $\bar{F}$, $m' = 2^n - m \leq 2^{n-1}$.

If for F and with respect to a variable x_i, $c_i \geq m/2$, then for F and with respect to the negated variable $\bar{x}_i$, $c'_i = m - c_i \leq m/2$.

When a function F can be expressed as $F = x_i G$ or $\bar{x}_i G$, where G is not dependent upon x_i, variable x_i is called a *nonessential variable*. If $F = \bar{x}_i G$, then $c_i = 0$. If $F = x_i G$, then $c_i = m(F)$. But with respect to the negated variable $\bar{x}_i$, $c'_i = m - c_i = 0$.

If a function F can be expressed as $F = x_i + G$ or $\bar{x}_i + G$, the complementary function is $\bar{F} = \bar{x}_i \bar{G}$ or $x_i \bar{G}$, and c_i or c'_i becomes 0 again.

DEFINITION. A Boolean function F of n variables is said to be *prime* if the following inequalities on n, $m(F)$, and the components of vector $C(F)$ hold.

$$2^{n-2} \geq m/2 \geq c_1 \geq c_2 \geq \cdots \geq c_n \geq 1$$

It can be shown that for a threshold function F of n variables, the complementation of variables, the permutation of variables, and/or the complementation of the function preserve the 1-realizability. It can also be shown that a function F is 1-realizable if and only if the function G after the elimination of nonessential variables is 1-realizable.

When a function is nonprime, it can be reduced to a prime function, and the original function is 1-realizable if and only if the resultant prime function is 1-realizable. Therefore only the prime functions need be considered.

The prime threshold functions of exactly 5 variables, in terms of Chow parameters, are listed in Table II.

TABLE II

Prime threshold functions of 5 variables.

m,	vector C					m,	vector C					m,	vector C				
6,	1	1	1	1	1	12,	5	4	3	3	2	14,	6	6	5	5	1
7,	2	2	1	1	1	12,	5	5	4	3	1	15,	5	5	5	4	4
8,	3	2	2	1	1	13,	5	4	4	4	2	15,	6	6	5	4	3
9,	3	3	3	1	1	13,	5	5	3	3	3	15,	7	6	5	5	2
9,	4	2	2	2	1	13,	5	5	5	4	1	15,	7	6	6	6	1
10,	4	3	3	2	1	13,	6	5	4	3	2	15,	7	7	4	4	3
10,	4	4	4	1	1	13,	6	6	4	4	1	16,	5	5	5	5	5
11,	4	4	3	3	1	14,	5	5	4	4	3	16,	6	6	6	4	4
11,	5	3	3	2	2	14,	6	6	4	3	3	16,	7	7	5	5	3
11,	5	4	4	2	1	14,	6	5	5	4	2	16,	7	7	7	7	1
12,	4	4	4	4	1												

3.6 MODIFIED CHOW PARAMETERS

It is easily seen that if a prime function F is 1-realizable, the weights are all negative. Now let us consider the complementary function $\bar{F}$.

For $\bar{F}$,

$$m' = 2^n - m, \text{ or } m'/2 = 2^{n-1} - m/2$$

and

$$c_i' = 2^{n-1} - c_i$$

Therefore

$$2^{n-1} \leq m'/2 \leq c_1' \leq c_2' \leq \dots \leq c_n' \leq 2^{n-1} - 1 \qquad (3.6\text{–}1)$$

Thus in the realization of $\bar{F}$, the weights will be all positive.
We shall call c_1', c_2', ..., c'_n and m' the *modified Chow parameters.*

3.7 DERTOUZOS' CHARACTERIZATION

Dertouzos[2,3] developed an approach based on correlation technique for the analysis and synthesis of threshold logic elements. The realizability of Boolean functions of n variables is shown to be dependent on a $(n+1)$-tuple of real numbers, called the *characterization vector*, which is derivable solely from the given Boolean function.

In his development, binary variables and functions assume a value of either -1 or $+1$.

Let $y_i \in \{0, 1\}$

$x_i \in \{-1, +1\}$

$F'(y_1, \dots, y_n)$ be a threshold function in n variables $\{y_i\}$ and $\in \{0, 1\}$.

$F(x_1, \dots, x_n)$ be the corresponding threshold function in n variables $\{x_i\}$ and $\in \{-1, 1\}$.

$$R[F'] = (a_1', \dots, a_n'; T')$$

$$R[F] = (a_1, \dots, a_n; T)$$

The transformation from y_i, F', a_i', and T' to x_i, F, a_i, and T are as follows:

$$x_i = 2y_i - 1$$

$$F = 2F' - 1$$

$$a_i = a_i'$$

$$T = 2T' - \sum_{i=1}^{n} a_i.$$

Let $X_j = (x_1, \dots, x_n)$ be the input vector or vertex whose decimal equivalent is j when $x_i \in \{0, 1\}$.

Then the defining inequalities of a threshold function become

$$F(X_j) = 1, \qquad \text{if } f_A(X_j) = \sum_{i=1}^{n} a_i x_i \geq T$$
$$F(X_j) = -1, \quad \text{if } f_A(X_j) = \sum_{i=1}^{n} a_i x_i < T \tag{3.7–1}$$

Considering it as a strict separating system and shifting T to the left-hand side relabeled as $-a_0$, we have

$$F(X_j) = 1, \qquad \text{if } f_A(X_j) = \sum_{i=1}^{n} a_i x_i + a_0 > 0$$
$$F(X_j) = -1, \quad \text{if } f_A(X_j) = \sum_{i=1}^{n} a_i x_i + a_0 < 0 \tag{3.7–2}$$

For convenience define

$$\langle G(j) \rangle \triangleq \sum_{j=0}^{2^n-1} G(j) \tag{3.7–3}$$

Then we have

$$\langle x_{ji} \rangle = 0 \tag{3.7–4}$$

$$\langle x_{ji_1} x_{ji_2} \rangle = 2^n \delta_{i_1 i_2} \tag{3.7–5}$$

where $\delta_{i_1 i_2}$ = Kronecker delta.

Equations (3.7–2) permit the following contraction without loss in either necessity or sufficiency for realizability.

$$f_A(X_j)F(X_j) = |f_A(X_j)|, f_A(X_j) \neq 0 \text{ for all } j \tag{3.7–6}$$

Summation over all j on both sides of Equation (3.7–6) yields

$$\langle f_A(X_j)F(X_j) \rangle = \langle |f_A(X_j)| \rangle, f_A(X_j) \neq 0 \text{ for all } j \tag{3.7–7}$$

Substitution of $f_A(X_j)$ from (3.7–2) into (3.7–7) and interchange of the order of summations over j and i yield

$$\sum_{i=1}^{n} a_i \langle x_{ji} F(X_j) \rangle + a_0 \langle F(X_j) \rangle$$
$$= \langle | f_A(X_j) | \rangle \triangleq \langle | \sum_{i=1}^{n} a_i x_{ji} + a_0 | \rangle \tag{3.7–8}$$

The quantities $\langle x_{ji} F(X_j) \rangle$ and $\langle F(X_j) \rangle$ are dependent solely on the given Boolean function $F(X_j)$ and the input variables x_{ji}, and can be readily obtained upon specification of $F(X_j)$.

Now define $b_i = \langle x_{ji}F(X_j)\rangle$, $i = 1, \ldots, n$

$$b_0 = \langle F(X_j)\rangle \qquad (3.7\text{–}9)$$

$B = (b_1, \ldots, b_n;\, b_0)$ is called by Dertouzos the *characteristic vector* of $F(X_j)$.

Substitution of (3.7–9) into (3.7–8) yields

$$\sum_{i=0}^{n} a_i b_i = \langle |\sum_{i=1}^{n} a_i x_{ji} + a_0|\rangle = \langle |f_A(X_j)|\rangle, f_A(X_j) \neq 0 \quad \text{for all } j \qquad (3.7\text{–}10)$$

Equation (3.7–10) constitutes a necessary and sufficient condition for 1-realizability of a Boolean function.

THEOREM 3.7.1. A given Boolean function of n variables is 1-realizable iff there exists a weight-threshold vector $A = (a_1, a_2, \ldots, a_n;\, -T)$ such that

$$A \cdot B = \langle |f_A(X_j)|\rangle$$

3.8 COMPARISON OF CHOW'S AND DERTOUZOS' CHARACTERIZATIONS

It is now evident that every threshold function F is completely characterized by a vector of $(n+1)$ components, as Chow and Dertouzos developed. Suppose we consider all positive weights. The modified Chow parameters m' and C' are related to Dertouzos' b_i as follows:

$$b_0 = 2m' - 2^n$$

$$b_i = 4c_i' - 2m', \quad i = 1, 2, \ldots, n.$$

Insofar as threshold functions are concerned, either Chow's parameters or b_i contain equally all the necessary information. However, in the tabulation of threshold functions, the use of the b_i's provides a more economical representation.

REFERENCES

1. Chow, C. K. (1961). On the characterization of threshold functions, in "Switching Circuit Theory and Logical Design", *A.I.E.E. Special Publication* **S134,** 34–38.
2. Dertouzos, M. L. (1964). An approach to single-threshold-element synthesis, *I.E.E.E. Trans. electronic Comput.* EC**13,** 519–528.
3. Dertouzos, M. L. (1965). "Threshold Logic", M.I.T. Press, U.S.A.
4. Hu, S. T. (1965). "Threshold Logic", University of California Press, U.S.A.
5. McNaughton, R. (1961). Unate truth functions, *I.R.E. Trans. electronic Comput.* EC**10,** 1–6.
6. Winder, R. O. (1962). "Threshold Logic", Doctoral dissertation, Mathematics Department, Princeton University.

CHAPTER 4

Unateness, Monotonicity, and Asummability

4.1 POSITIVITY

In this chapter we shall study the most important properties of threshold functions, *viz.* unateness, monotonicity, and asummability.

Positivity and unateness have been studied extensively.[4,5,8,9,10,12,13,14] For a threshold function unateness is a relatively weak necessary condition. However, since unateness can be easily tested for from the function itself, it serves very well as a first step in the testing and realization of threshold functions.

A switching function F of n variables is said to be *positive* in a variable x_i iff F has a normal form in which no term has $\bar{x}_i$ as a factor. Similarly, F is said to be *negative* in x_i iff F has a normal form in which no term has x_i as a factor.

For instance, $F = x_1\bar{x}_2+x_1x_3x_4+\bar{x}_2\bar{x}_3\bar{x}_4$ is positive in x_1, is negative in x_2, and is neither positive nor negative in x_3 and x_4.

When F is positive in x_i, it is negative in $\bar{x}_i$. When F is negative in x_i, it is positive in $\bar{x}_i$.

This simply means that, for a function F positive in x_i, if it contains a minterm $x_1^* \ldots x_{i-1}^* \; \bar{x}_i x_{i+1}^* \ldots x_n^*$, it must also contain the minterm $x_1^* \ldots x_{i-1}^* x_i x_{i+1}^* \ldots x_n^*$, so that these two minterms can form the simple product $x_1^* \ldots x_{i-1}^* x_{i+1}^* \ldots x_n^*$ which does not have $\bar{x}_i$ as a factor. In geometrical interpretation, if F contains a vertex in the $(n-1)$-cube $\bar{x}_i$, F must also contain the corresponding vertex in the $(n-1)$-cube x_i, which differs from the vertex in the $(n-1)$-cube $\bar{x}_i$ only in the ith coordinate. These two vertices are obviously consecutive and form a 1-cube.

Let F be expanded along the variable x_i, i.e.

$$F = x_i F_{x_i} + \bar{x}_i F_{\bar{x}_i}$$

According to definition of positivity, and the discussion on reduced functions in Section 3.2,

$$F_{x_i} \supset F_{\bar{x}_i}, \text{ if } F \text{ is positive in } x_i;$$

$$F_{x_i} \subset F_{\bar{x}_i}, \text{ if } F \text{ is negative in } x_i.$$

If F is both positive and negative in x_i, then

$$F_{x_i} \supset F_{\bar{x}_i} \text{ and } F_{x_i} \subset F_{\bar{x}_i},$$

or

$$F_{x_i} = F_{\bar{x}_i}.$$

Therefore

$$F = x_i F_{x_i} + \bar{x}_i F_{\bar{x}_i}$$

$$= F_{x_i},$$

or F is independent of x_i.

THEOREM 4.1.1. If a switching function F of n variables is positive in x_i, it can be expressed as $x_i G_1 + G_2$, where G_1 and G_2 are functions independent of x_i, such that $G_1 \supset G_2$.

PROOF: $F = x_i F_{x_i} + \bar{x}_i F_{\bar{x}_i}$

Since $F_{x_i} \supset F_{\bar{x}_i}$,

$$F = x_i F_{x_i} + x_i F_{\bar{x}_i} + \bar{x}_i F_{\bar{x}_i}$$

$$= x_i F_{x_i} + F_{\bar{x}_i}$$

By setting $G_1 = F_{x_i}$ and $G_2 = F_{\bar{x}_i}$, we have

$$F = x_i G_1 + G_2$$

Q.E.D.

COROLLARY 4.1.1–1. If a switching function F of n variables is positive in x_i and not negative in x_i, then $c_i > m/2$.

COROLLARY 4.1.1–2. If a switching function F of n variables is independent of x_i, then $c_i = m/2$.

4.2 UNATE FUNCTIONS

A switching function F of n variables is said to be *unate in* x_i iff F is either positive or negative in x_i.

A switching function F of n variables is said to be *unate* iff it is unate in each of its variables.

For instance, $F = x_1\bar{x}_2 + x_3\bar{x}_4$ is a unate function, because it is positive in x_1 and x_3, and is negative in x_2 and x_4.

Unate functions are also a restricted class of switching functions, which is larger than the class of threshold functions and contains the latter class, as will be shown later. In the study of threshold functions unate functions play an important role because it is relatively easy to determine whether a given function is a unate function or not, while the testing of a threshold function is rather involved, especially when n is large.

Two switching functions F_1 and F_2 are said to be *comparable* iff either

$F_1 \supset F_2$ or $F_1 \subset F_2$. The property of being comparable is called *comparability*.

It is obvious that the reduced functions F_{x_i} and $F_{\bar{x}_i}$ of a unate function F expanded along any variable x_i are comparable.

THEOREM 4.2.1. A unate function F of n variables is positive in x_i and not negative in x_i iff $c_i > m/2$.

PROOF: Necessity is obvious and is true for an arbitrary switching function, as given by Corollary 4.1.1–1.

Sufficiency.

$F = x_i F_{x_i} + \bar{x}_i F_{\bar{x}_i}$

$m(F_{x_i}) = c_i$, $m(F_{\bar{x}_i}) = m - c_i$

Since

$c_i > m/2$,

$m(F_{x_i}) > m(F_{\bar{x}_i})$

Hence $F_{x_i} \not\subset F_{\bar{x}_i}$ and F is not negative in x_i. But F is a unate function, and must be either positive or negative in x_i. Therefore F is positive in x_i. Q.E.D.

THEOREM 4.2.2. A unate function F of n variables is independent of x_i iff $x_i = m/2$.

PROOF: Necessity is obvious and is true for an arbitrary switching function, as given by Corollary 4.1.1–2.

Sufficiency. Since F is a unate function, F is either positive or negative in x_i, and consequently either

$F_{x_i} \supset F_{\bar{x}_i}$, or $F_{x_i} \subset F_{\bar{x}_i}$.

Since $c_i = m/2$,

$m(F_{x_i}) = m(F_{\bar{x}_i})$

Hence for either case,

$F_{x_i} = F_{\bar{x}_i}$,

or F is independent of x_i. Q.E.D.

THEOREM 4.2.3. If x_i occurs in a prime implicant of a switching function F of n variables, then F is not independent of x_i.

PROOF: Let the prime implicant subsuming x_i be $x_i Y$, where Y is the product of a subset of the remaining $n-1$ literals. Obviously $x_i Y$ is an r-cube. Suppose F is independent of x_i. F must also contain the r-cube $\bar{x}_i Y$, because for every vertex in $(n-1)$-cube x_i, there must exist a corresponding vertex in $(n-1)$-cube $\bar{x}_i$, differing from the former only in the coordinate x_i.

Thus F must contain the prime implicant $Y = x_i Y + \bar{x}_i Y$, which is an $(r+1)$-cube. Hence $x_i Y$ implies Y and is no longer a prime implicant, which is a contradiction. Therefore F is not independent of x_i.

Q.E.D.

COROLLARY 4.2.1–1. If x_i occurs in a prime implicant of a unate function F of n variables, then F is positive in x_i, and $\bar{x}_i$ cannot occur in any prime implicant of F.

COROLLARY 4.2.1–2. The disjunction of any subset of the set of prime implicants of a unate function F of n variables is unate.

COROLLARY 4.2.1–3. All the prime implicants of a unate function F of n variables, considered as r-cubes, contain at least one vertex in common.

If F is positive in $x_1, \ldots, x_n$, then obviously $x_1 \ldots x_n$ is a common vertex.

THEOREM 4.2.4. If a prime implicant P_1 of a switching function F of n variables implies the disjunction of several other prime implicants $P_2+P_3+\ldots+P_m = G$ of F, then G cannot be unate.

PROOF: Let $P_1 = x_1 \ldots x_i$ be a $(n-1)$-cube. Each prime implicant in G, i.e. $P_2, \ldots, P_m$, must have at least one literal x_j^*, $j = i+1, \ldots, n$, as a factor, because otherwise P_k, $k = 2, \ldots, m$, will either equal to P_1 or contain P_1. If $P_k = P_1$, the case is trivial and is not considered. If $P_k \supset P_1$, then P_1 implies P_k and P_1 is no longer a prime implicant, a contradiction to the assumption.

Now suppose G is unate, and positive in the remaining variables $x_{i+1}, \ldots, x_n$. Consider the vertex $x_1 \ldots x_i \bar{x}_{i+1} \ldots \bar{x}_n$. It is contained in P_1, but is not contained in G. Therefore $G \not\supset P_1$, or P_1 does not imply G, a contradiction to the assumption. Hence G is not unate. Q.E.D.

THEOREM 4.2.5. All prime implicants of a unate function F of n variables are essential prime implicants.

PROOF: Suppose a prime implicant P is not essential. Then it implies G, the disjunction of some other prime implicants of F. By Theorem 4.2.4, G cannot be unate. But by Corollary 4.2.3–2, G is unate. This is a contradiction. Hence P is an essential prime implicant. Q.E.D.

COROLLARY 4.2.5–1. A unate function F of n variables has a unique irredundant normal disjunctive form, which consists of the disjunction of all the prime implicants of F. Furthermore, this expression is unate, and so demonstrates the unateness of the function.

4.3 MONOTONICITY

Monotonicity and its related properties have also been studied intensively, especially by Winder.[11, 12, 13, 14] In increasing degree of restriction, it varies from 1-monotonicity, which is equivalent to unateness, to $[\frac{n}{2}]$ or complete monotonicity, which is equivalent to 2-asummability, a condition very close to linear separability.

A switching function F of n variables is said to be 1-*monotonic*, iff the reduced functions of F expanded along any one variable x_i, i.e. F_{x_i} and $F_{\bar{x}_i}$, are comparable.

A switching function F of n variables is said to be *k-monotonic*, iff the reduced functions of F expanded along any i variables, $1 \leq i \leq k$, are comparable.

A switching function F of n variables is said to be *completely monotonic* iff it is n-monotonic.

By definition, it is obvious that if a switching function is k-monotonic, it is also i-monotonic, $1 \leq i < k$.

THEOREM 4.3.1. A switching function F of n variables is unate iff it is 1-monotonic.

PROOF: This is obvious from the definitions of positivity, unateness, and 1-monotonicity. Q.E.D.

THEOREM 4.3.2. Let F be a 2-monotonic switching function of n variables, and $F_{x_j\bar{x}_k}$ and $F_{\bar{x}_jx_k}$ be two reduced functions of F expanded along two arbitrary variables x_j and x_k. Then

$$F_{x_j\bar{x}_k} \subset F_{\bar{x}_jx_k} \text{ and } F_{x_j\bar{x}_k} \not\supset F_{\bar{x}_jx_k} \text{ iff } c_j < c_k,$$

$$F_{x_j\bar{x}_k} = F_{\bar{x}_jx_k} \text{ iff } c_j = c_k.$$

PROOF:

$$c_j = m(F_{x_j}) = m(F_{x_jx_k}) + m(F_{x_j\bar{x}_k})$$

$$c_k = (F_{x_k}) = m(F_{x_jx_k}) + m(F_{\bar{x}_jx_k})$$

Hence

$$m(F_{x_j\bar{x}_k}) < m(F_{\bar{x}_jx_k}) \text{ iff } c_j < c_k$$

$$m(F_{x_j\bar{x}_k}) = m(F_{\bar{x}_jx_k}) \text{ iff } c_j = c_k$$

Since F is 2-monotonic, $F_{x_j\bar{x}_k}$ and $F_{\bar{x}_jx_k}$ are comparable.
Hence

$$F_{x_j\bar{x}_k} \subset F_{\bar{x}_jx_k} \text{ and } F_{x_j\bar{x}_k} \not\supset F_{\bar{x}_jx_k} \text{ iff } m(F_{x_j\bar{x}_k}) < m(F_{\bar{x}_jx_k}),$$
$$\text{and consequently iff } c_j < c_k;$$

$$F_{x_j\bar{x}_k} = F_{\bar{x}_jx_k} \text{ iff } m(F_{x_j\bar{x}_k}) = m(F_{\bar{x}_jx_k}),$$
$$\text{and consequently iff } c_j = c_k$$

Q.E.D.

DEFINITION. Two threshold functions of n variables, F_1 and F_2, are said to be *isobaric*[4] iff they can be realized with the same weight vector $A = (a_1, \ldots, a_n)$ and two different threshold values T_1 and T_2, respectively.

DEFINITION. m threshold functions of n variables, $F_1, F_2, \ldots, F_m$ are said to be isobaric iff they can be realized with the same weight vector $A = (a_1, \ldots, a_n)$ and m different threshold values $T_1, T_2, \ldots, T_m$, respectively.

THEOREM 4.3.3. Two isobaric threshold functions of n variables, F_1 and F_2, are comparable, and $F_1 \supset F_2$ if $T_1 \leq T_2$.

PROOF: $F_2 = 1$ iff $\sum_{i=1}^{n} a_i x_i > T_2$

Assume $T_1 \leq T_2$

Then $\sum_{i=1}^{n} a_i x_i \geq T_1$ iff $\sum_{i=1}^{n} a_i x_i \geq T_2$

Hence $F_1 = 1$ if $F_2 = 1$

Or $F_1 \supset F_2$ Q.E.D.

COROLLARY 4.3.3–1. Let $F_1, F_2, \ldots, F_m$ be m isobaric threshold functions with threshold values $T_1, T_2, \ldots, T_m$, respectively. If $T_1 \leq T_2 \leq \ldots \leq T_m$, then $F_1 \supset F_2 \supset \ldots \supset F_m$.

THEOREM 4.3.4. The reduced functions of a threshold function F of n variables, expanded along any one subset of variables, are isobaric threshold functions.

PROOF: Let F be expanded along m variables $x_1, \ldots, x_m$, and let $y_1 \ldots y_m$ and $z_1 \ldots z_m$ represent any two $(n-m)$-cubes. We shall show that the reduced functions for these two $(n-m)$-cubes, $F_{y_1 \ldots y_m}$ and $F_{z_1 \ldots z_m}$ are isobaric threshold functions.

For any vertex in the $(n-m)$-cube $y_1 \ldots y_m$, $F = F_{y_1 \ldots y_m}$. Hence $F_{y_1 \ldots y_m}$ can be realized by the same realization as F by fixing $y_1 = y_2 = \ldots = y_m = 1$.

Let $R[F] = (a_1, \ldots, a_n; T)$

Since $\sum_{i=1}^{m} a_i x_i + \sum_{j=m+1}^{n} a_j x_j \geq T$ if $F = 1$

$$\sum_{j=m+1}^{n} a_j x_j \geq T - \sum_{i=1}^{m} a_i x_i \big| y_1, \ldots, y_m = 1$$

$$= T' \qquad \text{if } F_{y_1 \ldots y_m} = 1$$

And since

$$\sum_{i=1}^{m} a_i x_i + \sum_{j=m+1}^{n} a_j x_j < T \quad \text{if } F = 0$$

$$\sum_{i=m+1}^{n} a_i x_i < T - \sum_{i=1}^{m} a_i x_i \big| y_1, \ldots, y_m = 1$$

$$= T' \qquad \text{if } F_{y_1 \ldots y_m} = 0$$

Therefore $F_{y_1 \ldots y_m}$ is a threshold function of $n-m$ variables with $R[F_{y_1 \ldots y_m}] = (a_{m+1}, \ldots, a_n; T')$.

Similarly we can show that $F_{z_1 \ldots z_m}$ is a threshold function of $n-m$ variables with a realization $R[F_{z_1 \ldots z_m}] = (a_{m+1}, \ldots, a_n; T'')$,

where $T'' = T - \sum_{i=1}^{m} a_i x_i | z_1, \ldots, z_m = 1$.

Hence $F_{y_1 \ldots y_m}$ and $F_{z_1 \ldots z_m}$ are isobaric. Q.E.D.

COROLLARY 4.3.4–1. Any threshold function F of n variables is completely monotonic.

COROLLARY 4.3.4–2. Let $F_{y_1 \ldots y_m}$ and $F_{z_1 \ldots z_m}$ be two reduced functions of a threshold function F of n variables expanded along any one subset of variables $\{x_1, \ldots, x_m\}$. Then

$$F_{y_1 \ldots y_m} \supset F_{z_1 \ldots z_m} \text{ if } \sum_{i=1}^{m} a_i x_i | y_1, \ldots, y_m = 1 \geq \sum_{i=1}^{m} a_i x_i | z_1, \ldots, z_m = 1$$

THEOREM 4.3.5. An $[n/2]$-monotonic switching function F of n variables is completely monotonic, where $[n/2]$ is the greatest integer $\leq n/2$.

PROOF: Suppose F is not k-monotonic, $k > [n/2]$, but is j-monotonic, $j = 1, \ldots, k-1$. Then there exist reduced functions expanded along k variables, which are not comparable. Let the k variables be $x_1, \ldots, x_k$, and the two reduced functions which are not comparable be $F_{y_1 \ldots y_k}$ and $F_{z_1 \ldots z_k}$. Then there exist two vertices $Y = y_1 \ldots y_n$ and $Z = z_1 \ldots z_n$ in F such that

$$y_{k+1} \cdots y_n \in F_{y_1 \ldots y_k}$$

$$y_{k+1} \cdots y_n \notin F_{z_1 \ldots z_k}$$

$$z_{k+1} \cdots z_n \in F_{z_1 \ldots z_k}$$

$$z_{k+1} \cdots z_n \notin F_{y_1 \ldots y_k}$$

Next consider the two vertices

$$U = y_1 \cdots y_k z_{k+1} \cdots z_n \text{ and } V = z_1 \cdots z_k z_{k+1} \cdots z_n.$$

Obviously,

$$U \notin y_1 \cdots y_k F_{y_1 \ldots y_k}$$

$$U \notin F$$

$$V \notin z_1 \cdots z_k F_{z_1 \ldots z_k}$$

$$V \notin F$$

Now since $Y \in F, y_1 \dots y_k \in F_{y_{k+1} \dots y_n}$

$$Z \in F, z_1 \dots z_k \in F_{z_{k+1} \dots z_n}$$

$$U \notin F, y_1 \dots y_k \notin F_{z_{k+1} \dots z_n}$$

$$V \notin F, z_1 \dots z_k \notin F_{y_{k+1} \dots y_n}$$

Hence $F_{y_{k+1} \dots y_n} \not\supset F_{z_{k+1} \dots z_n}$

and $F_{y_{k+1} \dots y_n} \not\subset F_{z_{k+1} \dots z_n}$,

or $F_{y_{k+1} \dots y_n}$ and $F_{z_{k+1} \dots z_n}$ are not comparable, which means that F is not $(n-k)$-monotonic.

Since $k > [n/2]$, $n-k \leq n-[n/2]-1 \leq [n/2]$. This contradicts the assumption that F is $[n/2]$-monotonic. Therefore F is completely monotonic. Q.E.D.

COROLLARY 4.3.5–1. All unate functions of three or fewer variables are threshold functions.

COROLLARY 4.3.5–2. All 2-monotonic functions of five or fewer variables are threshold functions.

THEOREM 4.3.6. If two switching functions F_1 and F_2 of n variables are comparable, then the reduced functions of F_1 and F_2 for any common $(n-m)$-cube $x_{i_1}^* \dots x_{i_m}^*$, i.e. $F_{1x_{i_1}^* \dots x_{i_m}^*}$ and $F_{2x_{i_1}^* \dots x_{i_m}^*}$, are comparable, for $m = 1, \dots, n$.

PROOF: We shall first show that the reduced functions expanded along a single variable x_i^*, i.e. $F_{1x_i^*}$ and $F_{2x_i^*}$, are comparable.

Suppose $F_{1x_i^*}$ and $F_{2x_i^*}$ are not comparable. Consider the $(n-1)$-cube. There exist a vertex $y_1 \dots y_{i-1} y_{i+1} \dots y_n$ of $F_{1x_i^*}$ which is not a vertex of $F_{2x_i^*}$ and a vertex $z_1 \dots z_{i-1} z_{i+1} \dots z_n$ of $F_{2x_i^*}$ which is not a vertex of $F_{1x_i^*}$. Now consider the n-cube. Obviously $y_1 \dots y_{i-1} x_i^* y_{i+1} \dots y_n$ is a vertex of F_1 but is not a vertex of F_2, and $z_1 \dots z_{i-1} x_i^* z_{i+1} \dots z_n$ is a vertex of F_2 but is not a vertex of F_1. Therefore F_1 and F_2 are not comparable, a contradiction to the assumption. Hence $F_{1x_i^*}$ and $F_{2x_i^*}$ are comparable.

To obtain the reduced functions expanded along a subset of m variables, we can expand F_1 and F_2 along a single common variable at a time. Then after each expansion the corresponding reduced functions are comparable. Hence after expansion along m variables the corresponding reduced functions are still comparable. Q.E.D.

THEOREM 4.3.7. A k-monotonic switching function of n variables is $(k+1)$-monotonic iff for every pair of opposite $(n-k-1)$-cubes $x_{i_1}^* \dots$

$x^*_{i_{k+1}}$ and $\bar{x}^*_{i_1} \ldots \bar{x}^*_{i_{k+1}}$, the corresponding reduced functions $F_{x^*_{i_1} \ldots x^*_{i_{k+1}}}$ and $F_{\bar{x}^*_{i_1} \ldots \bar{x}^*_{i_{k+1}}}$ are comparable.

PROOF: The necessity of this theorem is obvious. To prove the sufficiency, we shall show that for any pair of $(n-k-1)$-cubes that are not opposite, the corresponding reduced functions are comparable.

For two $(n-k-1)$-cubes not opposite to each other, there is at least one fixed coordinate that is same for the two $(n-k-1)$-cubes. Let this coordinate be the $i_{k+1}{}^{\text{th}}$, and let the two $(n-k-1)$-cubes be $y_{i_1} \ldots y_{i_k} x_{i_{k+1}}$ and $z_{i_1} \ldots z_{i_k} x_{i_{k+1}}$. Now first expand F along the first k variables and next along $x_{i_{k+1}}$.

The reduced functions for the $(n-k-1)$-cubes $y_{i_1} \ldots y_{i_k} x_{i_{k+1}}$ and $z_{i_1} \ldots z_{i_k} x_{i_{k+1}}$ are

$$F_{y_{i_1} \ldots y_{i_k} y_{i_{k+1}}} = (F_{y_{i_1} \ldots y_{i_k}})_{y_{i_{k+1}}}$$

and $F_{z_{i_1} \ldots z_{i_k} x_{i_{k+1}}} = (F_{z_{i_1} \ldots z_{i_k}})_{x_{i_{k+1}}}$

Since F is k-monotonic, $F_{y_{i1} \ldots y_{ik}}$ and $(F_{z_{i_1} \ldots z_{i_k}})_{x_{i_k}}$ are comparable.

By Theorem 4.3.6, $(F_{y_{i_1} \ldots y_{i_k}})_{x_{i_{k+1}}}$ and $(F_{z_{i_1} \ldots z_{i_k}})_{x_{i_{k+1}}}$ are comparable.

Since the reduced functions of F expanded along any $k+1$ variables are comparable, F is $(k+1)$-monotonic. Q.E.D.

THEOREM 4.3.8. For fixed n, the following sequence of properties defines a strictly decreasing sequence of class of functions: 1-monotonicity, 2-monotonicity, ..., $[n/2]$-monotonicity = complete monotonicity.

PROOF: By definition, a $(k+1)$-monotonic function is also k-monotonic. But the reverse is not necessarily true, as can be easily shown by counter-examples. Q.E.D.

The following examples show this strictly decreasing sequence of class of functions, i.e. a k-monotonic function is not necessarily $(k+1)$-monotonic.

$F_1 = x_1x_2 + x_3x_4$

F_1 is 1-monotonic but not 2-monotonic.

$F_2 = x_1x_2 + x_1x_3 + x_1x_4x_5x_6 + x_2x_3x_4 + x_2x_3x_5 + x_2x_4x_5x_6$

F_2 is 2-monotonic but not 3-monotonic.

$F_3 = x_1x_2 + x_1x_3 + x_1x_4 + x_1x_5x_6x_7x_8 + x_2x_3x_4x_5 + x_2x_3x_4x_6 + x_2x_3x_4x_7 + x_2x_3x_5x_6x_7x_8$

F_3 is 3-monotonic but not 4-monotonic.

4.4 ASUMMABILITY

k-asummability, for all integral values of $k \geq 2$, is a necessary and sufficient condition for threshold functions. Although stated slightly differently, it is derived from the concept of disjoint convex hulls, as has been studied extensively.[1,3,6,7,13]

An arbitrary switching function F of n variables is said to be *k-summable*, k being an integer ≥ 2, iff, for some integer j, $2 \leq j \leq k$, there are j vertices $\{X_1, \ldots, X_j\}$ in F and j vertices $\{Y_1, \ldots, Y_j\}$ in $\bar{F}$, not necessarily distinct, such that

$$\sum_{i=1}^{j} X_i = \sum_{i=1}^{j} Y_i.$$

If F is not k-summable, then F is said to be *k-asummable.*

F is said to be *summable* iff it is k-summable for some $k \geq 2$. F is said to be *asummable* iff it is not summable for any k.

LEMMA 4.4.1. A point X is in the convex hull $H(F)$ of a switching function F of n variables, iff it can be expressed as the average of m vertices of F, but not necessarily distinct, i.e.

$$X = \frac{1}{m} \sum_{i=1}^{m} X_i$$

PROOF: *Necessity.* By the definition of convex hull,

$$X = c_i X_i + \ldots + c_p X_p \tag{4.4.1–1}$$

where $\{X_i\}$, $i = 1, \ldots, p$ are the whole set of vertices in F, $c_1, \ldots, c_p$ are nonnegative real numbers satisfying

$$\sum_{i=1}^{p} c_i = 1.$$

Since the rational numbers are everywhere dense in the space R of all real numbers, $c_1, \ldots, c_p$ can be considered as nonnegative rational numbers. Multiplying both sides of (4.4.1–1) by m, the lowest common multiple of the denominators of c_i's, we have

$$mX = c'_1 X_1 + \ldots + c'_p X_p$$

where $c'_1 = mc_i$ is a nonnegative integer.

Now dividing both sides by m, we have

$$X = \frac{1}{m} \sum_{i=1}^{p} c'_i X_i, \quad X_i \text{ all distinct,}$$

$$= \frac{1}{m} \sum_{i=1}^{m} X_i, \quad X_i \text{ not necessarily distinct.}$$

Sufficiency. Let the given point be

$$X = \frac{1}{m} \sum_{i=1}^{m} X_i, \tag{4.4.1–2}$$

let p be the number of distinct vertices in F, and let $m_1, \ldots, m_p$ be the multiplicity of each vertex in (4.4.1–2), respectively. Then

$$\sum_{j=1}^{p} m_j = m, \text{ and (4.4.1–2) becomes}$$

$$X = \frac{1}{m} \sum_{j=1}^{p} m_j X_j$$

$$= \sum_{j=1}^{p} \frac{m_j}{m} X_j$$

By definition, X is a point of $H(F)$. Q.E.D.

THEOREM 4.4.1. A switching function F of n variables is a threshold function iff it is asummable.

PROOF: *Necessity.* Assume that F is summable. There exist a point X in $H(F)$ and a point Y in $H(\bar{F})$ such that

$$X = \frac{1}{m} \sum_{i=1}^{m} X_i = \frac{1}{m} \sum_{i=1}^{m} Y_i = Y$$

Therefore $H(F)$ and $H(\bar{F})$ are not disjoint. By Theorem 2.3.1 F is not a threshold function, a contradiction.

Sufficiency. Assume that F is not a threshold function. Then $H(F)$ and $H(\bar{F})$ are not disjoint, and there is at least one point X common to $H(F)$ and $H(\bar{F})$, which can be expressed as

$$X = \frac{1}{m} \sum_{i=1}^{m} X_i = \frac{1}{m'} \sum_{i=1}^{m'} Y_i,$$

Or $m' \sum_{i=1}^{m} X_i = m \sum_{i=1}^{m'} Y_i$.

Considering the repeated vertices separately, we have

$$\sum_{j=1}^{mm'} X_j = \sum_{j=1}^{mm'} Y_j$$

Hence F is mm'-summable, a contradiction again. Q.E.D.

LEMMA 4.4.2. Let $F_{x_{i_1}^* \ldots x_{i_m}^*}$ and $\bar{F}_{x_{i_1}^* \ldots x_{i_m}^*}$ be respectively the reduced functions of F and $\bar{F}$, two mutual complementary switching functions of n variables, for any common $(n-m)$-cube $x_{i1}^* \ldots x_{im}^*$, $1 \le m \le n$. Then

$$\bar{F}_{x_{i_1}^* \ldots x_{i_m}^*} = \overline{(F_{x_{i_1}^* \ldots x_{i_m}^*})}$$

PROOF:

$$\begin{aligned} & F_{x_{i_1}^* \ldots x_{i_m}^*} + \bar{F}_{x_{i_1}^* \ldots x_{i_m}^*} \\ &= (F+\bar{F})_{x_{i_1}^* \ldots x_{i_m}^*} \\ &= 1_{x_{i_1}^* \ldots x_{i_m}^*} \\ &= 1 \end{aligned}$$

$$\begin{aligned} & F_{x_{i_1}^* \ldots x_{i_m}^*} \quad \bar{F}_{x_{i_1}^* \ldots x_{i_m}^*} \\ &= (F\bar{F})_{x_{i_1}^* \ldots x_{i_m}^*} \\ &= 0_{x_{i_1}^* \ldots x_{i_m}^*} \\ &= 0 \end{aligned}$$

Hence $\bar{F}_{x_{i_1}^* \ldots x_{i_m}^*} = \overline{(F_{x_{i_1}^* \ldots x_{i_m}^*})}$ Q.E.D.

THEOREM 4.4.2. A switching function F of n variables is completely monotonic iff it is 2-asummable.

PROOF: *Necessity.* Assume that F is 2-summable. There exist two vertices U and V in F and two vertices Y and Z in $\bar{F}$ such that

$$U+V = Y+Z$$

Or $u_i+v_i = y_i+z_i, \qquad i = 1, \ldots, n$

If $v_i = u_i$, then $y_i = z_i = u_i = v_i$.

If $v_i = \bar{u}_i$, then either $y_i = u_i$ and $z_i = v_i = \bar{u}_i$, or $y_i = v_i = \bar{u}_i$ and $z_i = u_i$.

Let us relabel the subscripts such that for $i = 1, \ldots, k$, $u_i = v_i = y_i = z_i$; for $i = k+1, \ldots, k+m$, $u_i = y_i$, $v_i = z_i = \bar{u}_i$; for $i = k+m+1, \ldots, n$, $u_i = z_i$, $v_i = y_i = \bar{u}_i$. Then we have

$$\begin{aligned} U &= u_1 \ldots u_k u_{k+1} \ldots u_{k+m} u_{k+m+1} \ldots u_n \\ V &= u_1 \ldots u_k \bar{u}_{k+1} \ldots \bar{u}_{k+m} \bar{u}_{k+m+1} \ldots \bar{u}_n \\ Y &= u_1 \ldots u_k u_{k+1} \ldots u_{k+m} \bar{u}_{k+m+1} \ldots \bar{u}_n \\ Z &= u_1 \ldots u_k \bar{u}_{k+1} \ldots \bar{u}_{k+m} u_{k+m+1} \ldots u_n \end{aligned}$$

Now consider the reduced functions $F_{u_1 \ldots u_k u_{k+1} \ldots u_{k+m}}$, $F_{u_1 \ldots u_k \bar{u}_{k+1} \ldots \bar{u}_{k+m}}$, $\bar{F}_{u_1 \ldots u_k u_{k+1} \ldots u_{k+m}}$, $\bar{F}_{u_1 \ldots u_k \bar{u}_{k+1} \ldots \bar{u}_{k+m}}$, and the $(n-k-m)$-space. It is obvious that $u_{k+m+1} \ldots u_n$ is a vertex of $F_{u_1 \ldots u_k u_{k+1} \ldots u_{k+m}}$ and also is a vertex of $\bar{F}_{u_1 \ldots u_k \bar{u}_{k+1} \ldots \bar{u}_{k+m}}$. By

LEMMA 4.4.2, it is not a vertex of $F_{u_1 \ldots u_k \bar{u}_{k+1} \ldots \bar{u}_{k+m}}$. Hence

$$F_{u_1 \ldots u_k u_{k+1} \ldots u_{k+m}} \not\subset F_{u_1 \ldots u_k \bar{u}_{k+1} \ldots \bar{u}_{k+m}}$$

Similarly, $\bar{u}_{k+m+1} \ldots \bar{u}_n$ is a vertex of $F_{u_1 \ldots u_k \bar{u}_{k+1} \ldots \bar{u}_{k+m}}$ and also a vertex of $\bar{F}_{u_1 \ldots u_k u_{k+1} \ldots u_{k+m}}$. By Lemma 4.4.2, it is not a vertex of $F_{u_1 \ldots u_k u_{k+1} \ldots u_{k+m}}$. Hence

$$F_{u_1 \ldots u_k u_{k+1} \ldots u_{k+m}} \not\supset F_{u_1 \ldots u_k \bar{u}_{k+1} \ldots \bar{u}_{k+m}}$$

Therefore F is not completely monotonic, a contradiction.

Sufficiency. Assume that F is not completely monotonic. There exist reduced functions $F_{y_1 \ldots y_m}$ and $F_{z_1 \ldots z_m}$ that are not comparable, where $y_1 \ldots y_m$ and $z_1 \ldots z_m$ are isomerous $(n-m)$-cubes. Let $y_{m+1} \ldots y_n$ be a vertex of $F_{y_1 \ldots y_m}$ but not a vertex of $F_{z_1 \ldots z_m}$, and $z_{m+1} \ldots z_n$ be a vertex of $F_{z_1 \ldots z_m}$ but not a vertex of $F_{y_1 \ldots y_m}$. Then $y_{m+1} \ldots y_n$ is a vertex of $\bar{F}_{z_1 \ldots z_m}$ and $z_{m+1} \ldots z_n$ is a vertex of $\bar{F}_{y_1 \ldots y_m}$.

Now consider the n-cube again.

$X_1 = y_1 \ldots y_m y_{m+1} \ldots y_n$ is a vertex of F.

$X_2 = z_1 \ldots z_m z_{m+1} \ldots z_n$ is a vertex of F.

$X_3 = y_1 \ldots y_m z_{m+1} \ldots z_n$ is a vertex of $\bar{F}$.

$X_4 = z_1 \ldots z_m y_{m+1} \ldots y_n$ is a vertex of $\bar{F}$.

Therefore $X_1 + X_2 = X_3 + X_4$ and F is 2-summable, a contradiction again. Q.E.D.

REFERENCES

1. Chow, C. K. (1961). Boolean functions realizable with single threshold devices, *Proc. Inst. Radio Engrs* **49,** 270–271.
2. Coates, C. L., and Lewis, P. M. II. (1961). Linearly separable switching functions, *J. Franklin Inst.* **272,** 360–410.
3. Elgot, C. C. (1961). Truth functions realizable by single threshold organs, in "Switching Circuit Theory and Logical Design", *A.I.E.E. Special Publication* **S134**, 225–245.
4. Ercoli, P., and Mercurio, L. (1963). Threshold logic with one or more than one threshold, *Proc. IFIP Congress 62*, North-Holland Publishing Co. 741–746.
5. Fischler, M. A. (1962). Investigations concerning the theory and synthesis of linearly separable switching functions, Ph.D. dissertation, Department of Electrical Engineering, Stanford University, U.S.A.
6. Gabelman, I. J. (1962). A note on the realization of Boolean function using a single threshold element, *Proc. Inst. Radio Engrs* **50,** 225–226.
7. Highleyman, H. (1961). A note on linear separation, *I.R.E. Trans. electronic Comput.* EC**10,** 777–778.

8. Hu, S. T. (1965). "Threshold Logic", University of California Press.
9. Liss, D. (1963). A test for unate functions, *I.E.E.E. Trans. electronic Comput.* EC**12,** 405.
10. McNaughton, R. (1961). Unate truth functions, *I.R.E. Trans. electronic Comput.* EC**10,** 1–6.
11. Muroga, S., Toda, I., and Takasu, S. (1961). Theory of majority decision elements, *J. Franklin Inst.* **271,** 376–418.
12. Paull, M. C., and McCluskey, E. J. Jr. (1960). Boolean functions realizable with single threshold devices, *Proc. Inst. Radio Engrs* **48,** 1335–1337.
13. Winder, R. O. (1961). Single state threshold logic, in "Switching Circuit Theory and Logical Design", *A.I.E.E. Special Publication* S**134,** 321–332.
14. Winder, R. O. (1962). Threshold Logic, Doctoral dissertation, Mathematics Department, Princeton University, U.S.A.

CHAPTER 5

Admissible Transformations, Partial Ordering, and Weights

5.1 ADMISSIBLE TRANSFORMATIONS OF VARIABLES AND FUNCTIONS

In this chapter we shall study the equivalence classes of switching functions as obtained by admissible transformations.[5,6,11] We shall also study canonical switching functions, partial ordering of vertices, assignment of weights to the variables, and the properties of threshold functions related to weights.[1,4,6,7,8,9,10,12]

An *admissible transformation* of B^n is a bijective (one-to-one and onto) transformation $\alpha: B^n \to B^n$ which can be extended to a linear transformation of Euclidean space.

It can be shown that all admissible transformations correspond to *permuting* and *complementing* subsets of the n variables.

It can also be shown that the set of all admissible transformations of B^n forms a group with composition as the group operation. This group is called the *hyper-octahedral group* and is denoted by O_n. Since there are altogether 2^n possible ways of complementation and $n!$ possible ways of permutation, the hyper-octahedral group O_n is of order $2^n(n!)$.

Every admissible transformation $\alpha \in O_n$ induces a bijective transformation $\alpha^{\#}: \Phi \to \Phi$ of the set Φ of all switching functions, defined by

$$\alpha^{\#}(F) = F \cdot \alpha$$

for each $F \in \Phi$, where $F \cdot \alpha$ stands for the composition of α and F.

For instance, let $F = x_1 + \bar{x}_2 x_3$ and α be a permutation of the coordinates x_1 and x_2. Then $\alpha^{\#}(F) = x_2 + \bar{x}_1 x_3$.

It can be shown that the hyper-octahedral group O_n acts on Φ as a transformation group of Φ.

Two switching functions F and G of n variables are said to be *admissibly equivalent* iff there exists an admissible transformation $\alpha \in O_n$ such that

$$\alpha^{\#}(F) = F \cdot \alpha = G.$$

For instance, $F = x_1 + \bar{x}_2 x_3$ and $G = x_2 + \bar{x}_1 x_3$ are admissibly equivalent.

Admissible equivalence is thus an equivalence relation in the set of switching functions. The 2^{2^n} switching functions of n variables are divided into disjoint equivalence classes which are sometimes called the *symmetry types*.

It is beyond this book to discuss admissible transformations in detail. We shall, however, state some important properties of switching functions related to complementation and permutation without proof.

It can be shown that for any complementation and/or permutation of variables, i.e. any admissible transformation α, the transformation $\alpha^{\#}: \Phi \rightarrow \Phi$ preserves linear separability, k-summability, k-asummability, k-monotonicity, and complete monotonicity.

Let $F = F(x_1, x_2, \ldots, x_n)$

$$G = F(x_1, \ldots, x_{j-1}, \bar{x}_j, x_{j+1}, \ldots, x_n)$$

Then it can be easily shown that

$$G_{x_j} = F_{\bar{x}_j}, G_{\bar{x}_j} = F_{x_j}$$

For instance, let

$$F = x_1x_2 + x_2\bar{x}_3x_4 + \bar{x}_2x_3, \text{ and } G = x_1\bar{x}_2 + \bar{x}_2\bar{x}_3x_4 + x_2x_3,$$

then

$$F_{x_2} = x_1 + \bar{x}_3x_4, \text{ and } G_{\bar{x}_2} = x_1 + \bar{x}_3x_4.$$

If F is positive (negative) in x_j, then G is negative (positive) in x_j.

F is independent of x_j iff $F = G$, and consequently, $F_{x_j} = F_{\bar{x}_j}$.

As shown in Chapter 4, a unate function F of n variables is independent of x_j iff $c_j = m/2$.

It is easily seen that every unate function is admissibly equivalent to a positive switching function.

A switching function F of n variables is said to be *symmetric* in a subset of m variables, $2 \leq m \leq n$, iff F remains invariant under any permutation of the m variables.

For instance, $F = x_1x_2 + x_1x_3\bar{x}_4 + x_2x_3\bar{x}_4 + \bar{x}_1\bar{x}_2x_3x_4$ is symmetric in x_1 and x_2.

A switching function F of n variables is said to be a symmetric function iff F remains invariant under any permutation of the n variables.

For instance, $F = x_1x_2x_3 + x_1\bar{x}_2\bar{x}_3 + \bar{x}_1x_2\bar{x}_3 + \bar{x}_1\bar{x}_2x_3$ is a symmetric function, symmetric in x_1, x_2, and x_3.

It can be shown that a switching function F of n variables is symmetric in variables x_j and x_k if $F_{x_j\bar{x}_k} = F_{\bar{x}_jx_k}$.

A 2-monotonic function F of n variables is symmetric in x_j and x_k if $c_j = c_k$.

THEOREM 5.1.1. If a threshold function F of n variables $\{x_1, \ldots, x_n\}$ has a realization $R[F] = (a_1, \ldots, a_n; T)$, then the switching function $G(x_1, \ldots, x_n) = F(x_1, \ldots, x_{j-1}, \bar{x}_j, x_{j+1}, \ldots, x_n)$ is also a threshold function and has a realization $R[G] = (a_1', \ldots, a_n'; T')$ such that $a_i' = a_i$ for $i \neq j$, $a_i' = -a_j$ for $i = j$, and $T' = T - a_j$.

PROOF: That G is also a threshold function is a direct consequence of preservation of linear separability by any admissible transformation. So we shall show that

$(a'_1, \ldots, a'_n; T')$ is a realization of G.

Let $x'_i = x_i$ for $i \neq j$ and $x'_i = \bar{x}_i = 1 - x_i$ for $i = j$.

Then, since $G(x_1, \ldots, x_n) = F(x_1, \ldots, x_{j-1}, \bar{x}_j, x_{j+1}, \ldots, x_n)$

$$= F(x'_1, \ldots, x'_n),$$

$$G = 1 \text{ if } \sum_{i=1}^{n} a_i x'_i \geq T$$

or
$$\sum_{i=1}^{j-1} a_i x_i + \sum_{i=j+1}^{n} a_i x_i + a_j(1 - x_j) \geq T$$

or
$$\sum_{i=1}^{j-1} a_i x_i + \sum_{i=j+1}^{n} a_i x_i + (-a_j) x_j \geq T - a_j$$

or
$$\sum_{i=1}^{n} a'_i x_i \geq T'.$$

Similarly,

$$G = 0 \text{ if } \sum_{i=1}^{n} a'_i x_i < T'$$

Hence $(a'_1, \ldots, a'_n; T')$ is a realization of G. Q.E.D.

For instance, for $F = x_1 + x_2 x_3 + x_2 x_4$, there exists a realization $R[F] = (3, 2, 1, 1; 3)$. Then for $G = x_1 + \bar{x}_2 x_3 + \bar{x}_2 x_4$, there exists a realization $R[G] = (3, -2, 1, 1; 1)$.

COROLLARY 5.1.1–1. If a threshold function F of n variables $\{x_1, \ldots, x_n\}$ has a realization $R[F] = (a_1, \ldots, a_n; T)$, then the switching function G obtained from F by complementing a subset of variables $\{x_{i_1}, \ldots, x_{i_m}\}$ is also a threshold function and has a realization $R[G] = (a'_1, \ldots, a'_n; T')$ such that $a'_1 = a_i$ for $x_i \notin \{x_{i_1}, \ldots, x_{i_m}\}$, $a'_i = -a_i$ for $x_i \in \{x_{i_1}, \ldots, x_{i_m}\}$, and $T' = T - \sum_{j=1}^{m} a_{i_j}$.

THEOREM 5.1.2. If a threshold function F of n variables has a realization $R[F] = (a_1, \ldots, a_n; T)$, then the threshold function G obtained from F by an arbitrary permutation of the variables is $R[F]$ with a permutation of the weights same as that of the variables.

This theorem is obvious and its proof is omitted.

5.2 CANONICAL SWITCHING FUNCTIONS

A switching function F of n variables is said to be *canonical* iff its parameters $m, c_1, \ldots, c_n$ satisfy the following relations:

$$m/2 \leq c_1 \leq c_2 \leq \ldots \leq c_n \tag{5.2.1}$$

It is seen that (5.2.1) is same as (3.6.1) for the complementary function of Chow's prime function, except that the relations $2^{n-2} \leq m/2$ and $c_n \leq 2^{n-1}-1$ need not be satisfied. This simply means that a canonical function may contain any number of vertices and may have nonessential variables.

Thus every switching function can be changed into a canonical function by complementation and permutation of the variables, without complementation of the function and/or elimination of nonessential variables. In other words, we have the following theorem.

THEOREM 5.2.1. Every switching function of n variables is admissibly equivalent to at least one canonical switching function.

Thus every equivalent class or symmetric type of switching functions of n variables contains at least one, and possibly more than one canonical switching function.

For an arbitrary switching function, its admissibly equivalent canonical switching function is not unique. But for certain restricted classes of switching functions, the admissibly equivalent canonical switching function is unique.

THEOREM 5.2.2. Every 2-monotonic switching function F of n variables is admissibly equivalent to a unique canonical switching function.

PROOF: Suppose there exist two distinct admissibly equivalent canonical switching functions F and G, and $F \neq G$. G can be obtained from F by complementation and/or permutation of the variables. There must be at least either one complementation of a variable x_i of which F is not independent, or one permutation of two variables x_j and x_k in which F is not symmetrical.

Suppose there is a complementation of x_i.

$$c_i(G) \geq m(g)/2 = m(F)/2$$

$$\text{Since } c_i(F) \geq m(F)/2,$$

$$c_i(G) = m - c_i(F) \leq m(F)/2$$

Hence $c_i(G) - c_i(F) = m(F)/2$

But $F_{x_i} \neq F_{\bar{x}_i}$ because F is not independent of x_i.

Hence $F_{x_i} \not\supset F_{\bar{x}_i}$ and $F_{x_i} \not\subset F_{\bar{x}_i}$. This implies that F is not 1-monotonic, and consequently also not 2-monotonic, a contradiction.

Suppose there is a permutation of x_j and x_k. Assume $j < k$.

$$c_j(F) \leq c_k(F)$$

But $c_j(F) = c_k(G) \geq c_j(F) = c_k(F)$

Hence $c_j(F) = c_k(F)$

But $F_{x_j\bar{x}_k} \neq F_{\bar{x}_j x_k}$ because F is not symmetrical in x_j and x_k.

Hence $F_{x_j\bar{x}_k} \not\supset F_{\bar{x}_j x_k}$ and $F_{x_j\bar{x}_k} \not\subset F_{\bar{x}_j x_k}$. This implies that F is not 2-monotonic, a contradiction again. Q.E.D.

COROLLARY 5.2.2–1. Every threshold function F of n variables is admissibly equivalent to a unique canonical switching function.

Canonical 2-monotonic switching functions are called *regular functions.*

5.3 WEIGHTS ASSIGNED TO VARIABLES

Now we shall study the properties of weights assigned to the variables of a switching function F of n variables. We can arbitrarily assign weights to the variables even if F is not a threshold function. However, they are restricted if they have to meet certain requirements.

A weight vector $A = (a_1, \ldots, a_n)$ is said to be *nonnegative* iff $a_i \geq 0$, for $i = 1, \ldots, n$.

A weight vector $A = (a_1, \ldots, a_n)$ is said to be *canonical* iff

$$0 \leq a_1 \leq a_2 \leq \ldots \leq a_n$$

For two arbitrary vertices X and Y of the n-cube, we define a relation $\leq$ by setting

$$X \leq Y$$

iff $XA^t \leq YA^t$ for any A with a certain specified restriction. Usually we specify that A is a canonical weight vector.

For instance, for $n = 4$, if $X = x_1x_2x_3\bar{x}_4 = 1110$ and $Y = x_1\bar{x}_2x_3x_4 = 1011$, then $X \leq Y$ for any canonical weight vector A.

It is seen that $\leq$ for vertices is a *partial ordering*, and therefore is reflexive, antisymmetric, and transitive. For a canonical weight vector, this partial ordering is called *canonical partial ordering of vertices.*

For the literals in a minterm or vertex, we shall call the complementation of an unnegated variable x_i to the negated one $\bar{x}_i$ a *negative complementation*, and that of $\bar{x}_i$ to x_i a *positive complementation.* Also we shall call the permutation of two variables $\bar{x}_j$ and x_k, $j < k$, to $\bar{x}_k$ and x_j, a *negative permutation*, and that of x_j and $\bar{x}_k$ to x_k and $\bar{x}_j$, a *positive permutation.* The permutation of x_j and x_k to x_k and x_j, and that of $\bar{x}_j$ and $\bar{x}_k$ to $\bar{x}_k$ and $\bar{x}_j$, are called *neutral permutations.*

For instance, from $x_1\bar{x}_2\bar{x}_3x_4$ to $x_1x_2\bar{x}_3\bar{x}_4$ there is a positive complementation from $\bar{x}_2$ to x_2 and a negative complementation from x_4 to $\bar{x}_4$. Also, these two complementations can be considered as a negative permutation from $\bar{x}_2$ and x_4 to $\bar{x}_4$ and x_2.

THEOREM 5.3.1. Let X and Y be two arbitrary vertices in the n-cube. With the restriction that A is nonnegative, $X \leq Y$ iff Y can be obtained from X by a set of positive complementations.

PROOF: *Necessity.* Let $X = x_1^*, \ldots, x_n^*$. All vertices in the n-cube can be obtained from X by complementing a subset of literals in X. Suppose Y is not obtained from X by a set of positive complementations. Then there is at least one negative complementation of a literal $x_j^* = x_j$ to $\bar{x}_j$. By setting a_j greater than the sum of all the other weights, we have

$$XA^t \geq a_j > \sum_{i \neq j} a_i \geq YA^t,$$

Or

$$XA^t \nleq YA^t, \text{ a contradiction.}$$

The sufficiency of this theorem is obvious. Q.E.D.

For instance, if $X = x_1\bar{x}_2\bar{x}_3x_4$ and $Y = x_1\bar{x}_2x_3x_4$, then $X \leqslant Y$ for any nonnegative A, because Y can be obtained from X by a positive complementation from $\bar{x}_3$ to x_3.

THEOREM 5.3.2. Let X and Y be two arbitrary vertices in the n-cube. With the restriction that A is canonical, $X \leq Y$ iff Y can be obtained from X by a set of complementations and/or permutations such that there exists one decomposition in which all the complementations and/or permutations are positive.

PROOF: *Necessity.* Suppose Y is obtained from X not as specified in the theorem. Then in every decomposition there exists at least one negative complementation or one negative permutation. Since the composition of a positive complementation and a negative complementation of two literals results in a permutation, we shall say there exist negative complementations only when the number of net negative complementations, i.e. the number of negative complementations minus the number of positive complementations, is positive.

It is obvious that when there is a positive complementation of $\bar{x}_i$ to x_i, the value of XA^t is increased by a_i; when there is a negative complementation, the value of XA^t is increased by $-a_i$; when there is a positive permutation of x_j and $\bar{x}_k$ to $\bar{x}_j$ and x_k, the value of XA^t is increased by $a_k - a_j$; when there is a negative permutation of $\bar{x}_j$ and x_k to x_j and $\bar{x}_k$, the value of XA^t is increased by $-(a_k - a_j)$.

When there is one net negative complementation of x_i to $\bar{x}_i$, the decomposition of the permutations can be made such that for any permutation of the literals x_j^* and x_k^*, either $i < j < k$, or $j < k < i$. Then we can set a_i greater than the sum of all the increases due to positive permutation, or

$$a_i > \sum(a_k - a_j)$$

Hence

$$YA^t = XA^t - a_i + \sum(a_k - a_j) < XA^t$$

Or

$$XA^t \nleq YA^t, \text{ a contradiction.}$$

It is obvious that this is also true when there is more than one net negative complementation.

Next consider the case that there is no net negative complementation. Then there is at least one negative permutation which cannot be avoided by any decomposition. Let us arrange X and Y as two rows and take the permutations from right to left, as shown in the following example.

$$X = \bar{x}_1\bar{x}_2x_3\bar{x}_4x_5x_6x_7\bar{x}_8$$

$$Y = x_1x_2\bar{x}_3x_4\bar{x}_5\bar{x}_6\bar{x}_7x_8$$

In X and Y above, only those coordinates which are different for X and Y are shown. Those coordinates which are same for X and Y contribute the same amount of weight to XA^t and YA^t, and therefore need not be considered.

From right to left, if a column of positive complementation is encountered, we shall combine it with the first column of negative complementation, thus forming a positive permutation. Then proceed to the next column not taken previously. When a column, say the kth column, of negative complementation is encountered, then combine it with the first column of positive complementation to its left, say the jth column, to form a negative permutation. In the above example, for instance, the 8th column is a positive complementation and is combined with the 7th column to form a positive permutation. The 6th column is a negative complementation and is combined with the 4th column to form a negative permutation. Now the columns between the jth and the kth columns are all of negative complementation, and we can set $a_k - a_j$ greater than the sum of all the increases due to positive permutations, resulting in $YA^t < XA^t$, or $XA^t \not\leq YA^t$, which is a contradiction.

It is obvious that this rule is also true when there are more than one negative permutations.

Sufficiency. The complementation part is obvious. The permutation part is true because for a canonical weight vector, $a_j < a_k$. Q.E.D.

For instance, if $X = x_1\bar{x}_2x_3x_4x_5\bar{x}_6$ and $Y = x_1x_2x_3\bar{x}_4x_5x_6$, then $X \leq Y$ for any canonical A, because Y can be obtained from X by a positive complementation from $\bar{x}_2$ to x_2 and a positive permutation from x_4 and $\bar{x}_6$ to x_6 and $\bar{x}_4$.

THEOREM 5.3.3. A switching function F of n variables is a positive 1-monotonic function iff, with the restriction that the weight vector A is nonnegative, $X \leq Y$ implies that if X is a vertex of F, Y is also a vertex of F, and that if Y is a vertex of $\bar{F}$, X is also a vertex of $\bar{F}$.

PROOF: *Necessity*. Let $X = x_1^* \ldots \bar{x}_i \ldots x_n^*$ and $Y = x_1^* \ldots x_i \ldots x_n^*$ and $X \leq Y$. Suppose X is a vertex of F and Y is a vertex in $\bar{F}$. Now X is a vertex of $\bar{x}_i F_{\bar{x}_i}$ and Y is a vertex of $x_i F_{\bar{x}_i}$ but not a vertex of $x_i F_{x_i}$. Hence $F_{x_i} \not\supset F_{\bar{x}_i}$, and consequently F is not positive in x_i, a contradiction.

Sufficiency. Suppose F is not a positive 1-monotonic function. Then F is not positive in at least one of the variables, and there exist one vertex $X = x_1^* \ldots \bar{x}_i \ldots x_n^*$ in F and one vertex $Y = x_1^* \ldots x_i \ldots x_n^*$ in $\bar{F}$. By Theorem 5.3.1, $X \leq Y$. Therefore if X is a vertex of F, Y is also a vertex of F, a contradiction. Q.E.D.

THEOREM 5.3.4. A switching function F of n variables is a regular switching function iff, with the restriction that the weight vector A is canonical, $X \leqslant Y$ implies that if X is a vertex of F, Y is also a vertex of F, and that if Y is a vertex of $\bar{F}$, X is also a vertex of $\bar{F}$.

PROOF: *Necessity*. When Y is obtained from X by positive complementations, the proof of necessity in Theorem 5.3.3 also holds for this theorem. So we shall consider that case that Y is obtained from X by positive permutations. Let $X = x_1^* \ldots x_j \ldots \bar{x}_k \ldots x_n^*$ and $Y = x_1^* \ldots \bar{x}_j \ldots x_k \ldots x_n^*$, and $X \leq Y$. Suppose X is a vertex of F and Y is a vertex of $\bar{F}$. Now X is a vertex of $x_j \bar{x}_k F_{x_j \bar{x}_k}$, and Y is a vertex of $\bar{x}_j x_k F_{x_j \bar{x}_k}$ but not a vertex of $\bar{x}_j x_k F_{\bar{x}_j x_k}$. Hence $F_{x_j \bar{x}_k} \not\subset F_{\bar{x}_j x_k}$, and consequently F is not regular, a contradiction.

Sufficiency. When Y is obtained from X by positive complementations, the sufficiency is proved in Theorem 5.3.3 for positive 1-monotonic functions with the restriction that the weight vector A is nonnegative. Since a regular function is necessarily positive 1-monotonic and a canonical weight vector is necessarily nonnegative, this is also true for this theorem.

Now consider the case that Y is obtained from X by positive permutations. Suppose F is not regular. Then there are two variables x_j and x_k such that $F_{x_j \bar{x}_k} \not\subset F_{\bar{x}_j x_k}$. This implies that there exists a vertex $X = x_1^* \ldots x_j \ldots x_k \ldots x_n^*$ in $x_j \bar{x}_k F_{x_j \bar{x}_k}$ such that $Y = x_1^* \ldots \bar{x}_j \ldots x_k \ldots x_n^*$ is not a vertex of $\bar{x}_j x_k F_{\bar{x}_j x_k}$, and consequently not a vertex of F, a contradiction. Q.E.D.

Since a threshold function F of n variables is necessarily 2-monotonic, the "only if" part of Theorem 5.3.4 holds for a canonical threshold function too.

A realization $R[F] = (a_1, \ldots, a_n; T)$ of a threshold function F of n variables is said to be canonical if the weight vector A is canonical.

THEOREM 5.3.5. For an arbitrary threshold function F of n variables, the following relations among weights must be satisfied:

(1) If $c_i > m/2$, then $a_i > 0$.
(2) If $c_i < m/2$, then $a_i < 0$.
(3) If $c_i = m/2$, it is admissible that $a_i = 0$;
if $a_i = 0$, then $c_i = m/2$.

(4) If $c_j < c_k$, then $a_j < a_k$.
(5) If $c_j = c_k$, it is admissible that $a_j = a_k$;
if $a_j = a_k$, then $c_j = c_k$.

PROOF:

(1) If $c_i > m/2$, by Theorem 4.2.1, F is positive in x_i but not negative in x_i.

$$F_{x_i} \supset F_{\bar{x}_i} \quad \text{and} \quad F_{x_i} \not\subset F_{\bar{x}_i}$$

There exists at least one vertex $Y = x_1^* \ldots x_i \ldots x_n^*$ in F such that $X = x_1^* \ldots \bar{x}_i \ldots x_n^*$ is a vertex of $\bar{F}$. For a realization $R[F] = (a_1, \ldots, a_n; T)$,

$$YA^t \geq T$$
$$XA^t < T$$

Suppose $a_i = 0$. Then

$$XA^t = YA^t \geq T, \text{ a contradiction.}$$

Suppose $a_i < 0$. Then

$$XA^t = YA^t - (-a_i) > YA^t \geq T, \text{ a contradiction.}$$

(2) The proof is similar to (1) and is omitted.

(3) If $c_i = m/2$, by Theorem 4.2.2, F is independent of x_i and $F_{x_i} = F_{\bar{x}_i}$. Therefore $X = x_1^* \ldots \bar{x}_i \ldots x_n^*$ is a vertex of F iff $Y = x_1^* \ldots x_i \ldots x_n^*$ is a vertex of F, and X is a vertex of $\bar{F}$ iff Y is a vertex of $\bar{F}$. Now for $a_i = 0$, either

$$YA^t = XA^t \geq T, \quad \text{or} \quad YA^t = XA^t < T$$

and there exists no contradiction. Hence $a_i = 0$ is admissible.

Since $c_i > m/2$ implies that $a_i > 0$ and $c_i < m/2$ implies that $a_i < 0$, $a_i \not> 0$ and $a_i \not< 0$ implies that $c_i \not> m/2$ and $c_i \not< m/2$. In other words,

$$a_i = 0 \text{ implies } c_i = m/2.$$

(4) If $c_j < c_k$, by Theorem 4.3.2,

$$F_{x_j\bar{x}_k} \subset F_{\bar{x}_j x_k} \quad \text{and} \quad F_{x_j\bar{x}_k} \not\supset F_{\bar{x}_j x_k}$$

Hence there exists at least one vertex $Y = x_1^* \ldots \bar{x}_j \ldots x_k \ldots x_n^*$ in $\bar{x}_j x_k F_{\bar{x}_j x_k}$ and consequently in F such that $X = x_1^* \ldots x_j \ldots \bar{x}_k \ldots x_n^*$ is not a vertex of $x_j\bar{x}_k F_{x_j\bar{x}_k}$ and consequently not a vertex of F. Hence X must be a vertex of $\bar{F}$.

Suppose $a_j = a_k$. Then

$$XA^t = YA^t + a_j - a_k = YA^t \geq T, \text{ a contradiction.}$$

Suppose $a_j > a_k$. Then

$$XA^t = YA^t + a_j - a_k > YA^t \geq T, \text{ a contradiction.}$$

(5) If $c_j = c_k$, by Theorem 4.3.2,

$$F_{x_j\bar{x}_k} = F_{\bar{x}_j x_k} \text{ and } F \text{ is symmetrical in } x_j \text{ and } x_k.$$

Therefore $X = x_1^* \dots x_j \dots \bar{x}_k \dots x_n^*$ is a vertex of F if $Y = x_1^* \dots \bar{x}_j \dots x_k \dots x_n^*$ is a vertex of F, and X is a vertex of $\bar{F}$ if Y is a vertex of $\bar{F}$.

For $a_j = a_k$, either

$$YA^t = XA^t - a_j + a_k = XA^t \geq T$$

or

$$YA^t = XA^t < T$$

and there exists no contradiction. Hence $a_j = a_k$ is admissible.

Since $c_j < c_k$ implies that $a_j < a_k$ and $c_j > c_k$ implies that $a_j > a_k$, $a_j \nless a_k$ and $a_j \ngtr a_k$ implies that $c_j \nless c_k$ and $c_j \ngtr c_k$. In other words,

$$a_j = a_k \text{ implies } c_j = c_k. \qquad \text{Q.E.D.}$$

It is thus seen that for a canonical threshold function, the weight vector, and consequently the realization, is also canonical.

THEOREM 5.3.6. Let $R[F] = (a_1, \dots, a_n; T)$ be a realization of a positive threshold function F of n variables. F is symmetrical in x_j and x_k iff $(a_j + a_k)/2$ can be used in place of both a_j and a_k to give, again, a realization of F.

PROOF: *Necessity.* If F is symmetrical in x_j and x_k, by Theorem 5.3.5, it is admissible that $a_j' = a_k'$. We shall prove that $a_j' = a_k' = (a_j + a_k)/2$ is acceptable, or that $R[F] = (A'; T) = (a_1, \dots, a_j', \dots, a_k', \dots, a_n; T)$ is also a realization of F.

There are altogether four different kinds of vertices in regard to x_j and x_k.

$$\begin{aligned} U &= x_1^* \dots x_j \dots x_k \dots x_n^* \\ V &= x_1^* \dots \bar{x}_j \dots \bar{x}_k \dots x_n^* \\ Y &= x_1^* \dots x_j \dots \bar{x}_k \dots x_n^* \\ Z &= x_1^* \dots \bar{x}_j \dots x_k \dots x_n^* \end{aligned}$$

For vertices U and V,

$$UA'^T = UA^T, \quad \text{and} \quad VA'^T = VA^T$$

Hence the use of $(a_j + a_k)/2$ in place of a_j and a_k results in the same inequalities.

For vertices Y and Z, since F is symmetrical in x_j and x_k, Y is a vertex of F if Z is a vertex of F, and vice versa.

Assume that $a_j < a_k$. If X and Z are vertices of F,

$$YA'^T - YA^T = ZA'^T - YA^T = \frac{a_j + a_k}{2} - a_j = \frac{a_k - a_j}{2} > 0$$

Hence

$$YA'^T = ZA'^T > YA^T \geq T$$

If Y and Z are vertices of $\bar{F}$,

$$YA'^T - ZA^T = ZA'^T - ZA^T = \frac{a_j + a_k}{2} - a_k = \frac{a_j - a_k}{2} < 0$$

Hence

$$YA'^T = ZA'^T < ZA^T < T$$

Therefore for any vertex X,

$$XA'^T \geq T \quad \text{if} \quad XA^T \geq T$$

$$XA'^T < T \quad \text{if} \quad XA^T < T$$

And $R[F] = (a_1, \ldots, a'_j, \ldots, a'_k, \ldots, a_n; T)$ is a realization of F.

Sufficiency. If $a'_j = a'_k = (a_j + a_k)/2$, then $c_j = c_k$, and $F_{x_j \bar{x}_k} = F_{\bar{x}_j x_k}$. Hence F is symmetrical in x_j and x_k. Q.E.D.

A vector $Y = (y_1, \ldots, y_n)$ is said to *precede* another vector $Z = (z_1, \ldots, z_n)$, if $y_i \leq z_i$ $(i = 1, \ldots, n)$ and $y_j < z_j$ for some j $(1 \leq j \leq n)$. That Y precedes Z is denoted by $Y < Z$.

THEOREM 5.3.7. Let F and G be Boolean functions of n variables. If $m(F) = m(G)$, $c_i(F) < m(F)/2$ $(i = 1, \ldots, n)$, and $C(G)$ precedes $C(F)$, then F is not 1-realizable.

PROOF: Suppose F is 1-realizable and $R[F] = (a_1, \ldots, a_n; T)$. Since $c_i(F) < m(F)/2$, $a_i < 0$ for $i = 1, \ldots, n$.

Since $C(G)$ precedes $C(F)$, F and G are distinct. Hence $F\bar{G}$ and $\bar{F}G$ are both nonempty and have the same number of vertices.

By definition of threshold function,

$$\sum_{X \in F\bar{G}} XA^T > \sum_{Y \in \bar{F}G} YA^T$$

This implies that

$$C(F)A^T > C(G)A^T$$

or

$$\sum_{i=1}^{n} [c_i(F) - c_i(G)]a_i > 0 \qquad (5.3.7\text{–}1)$$

But $c_i(F) - c_i(G)$ is nonnegative for all i and is positive for some i

because $C(G)$ precedes $C(F)$, and $a_i < 0$ because $c_i < m/2$. Therefore,

$$\sum_{i=1}^{n} [c_i(F) - c_i(G)]a_i < 0 \tag{5.3.7–2}$$

Inequalities (5.3.7–1) and (5.3.7–2) are contradictory. Therefore F is not 1-realizable. Q.E.D.

5.4 PARTIAL ORDERING AND REALIZATION

The partial ordering of vertices can be very useful for the testing and realization of threshold functions. From Theorems 5.3.3 and 5.3.4 it is seen that we can determine whether a vertex Y is in F or $\bar{F}$ by its partial ordering with another vertex X which is known to be in F or $\bar{F}$. However, for a positive 1-monotonic function, this "correlation" applies to only a subset of the vertices, and holds only with the restriction that the weight vector A is nonnegative. For a canonical 2-monotonic function, there can be more restrictions on the weight vector A, i.e. A is canonical, and the "correlation" applies to a larger subset of the vertices. For a threshold function, this "correlation" can apply to the whole set of vertices, with still more restrictions on the weight vector A. However, these restrictions are not easily seen, because they are in the form of a set of inequalities, and are to be found out through the solution of the set of inequalities. Fortunately, the restrictions on the weights from a positive 1-monotonic function to a 2-monotonic function and finally to a threshold function are consistent, i.e. there are monotonically increasing restrictions on the weights. Therefore the testing and realization of a threshold function becomes a problem of finding the increasing restrictions of weights. If it is known that the given function is a positive 1-monotonic function, we can assume that the weight vector is nonnegative, and if it is known that the given function is a regular function, we can assume the weight vector to be canonical, without causing any inconsistency later. Thus the testing of 1-monotonicity and 2-monotonicity becomes a very convenient and important first step in the testing and realization of threshold functions, because the former is relatively easy and straightforward.

Note that when $c_i = m/2$, it is admissible that $a_i = 0$, but a_i is not necessarily zero, and that when $c_j = c_k$, it is admissible that $a_j = a_k$, but a_j is not necessarily equal to a_k. This is because in the realization of a threshold function, there is a finite gap length $g = T_u - T_l$, and there exists an infinite number of realizations. Suppose for the realization of a function independent of x_i and symmetrical in x_j and x_k, T is such that $T_l < T \leq T_u$. Then if a_i, a_j, or a_k is changed by a small amount ϵ such that $|\epsilon| \leq T_u - T$ and $|\epsilon| < T - T_l$, then the resulting realization is still valid.

A vertex $X \in F$ is said to be *minimal* iff, for any arbitrary vertex $Y \in F$, $Y \leq X$ implies $X = Y$.

A vertex $X \in F$ is said to be *maximal* iff, for any arbitrary vertex $Y \in F$, $X \leq Y$ implies $X = Y$.

Since by definition $X \leqslant Y$ if $XA^T \leqslant YA^T$, in the testing and realization of threshold functions we need only consider the set of minimal vertices of F and the set of maximal vertices of $\bar{F}$, thus reducing the number of inequalities. For a nonnegative weight vector A, there exists a *minimal set*, sometimes called the *lowest set* $\{L\}$, and a *maximal set*, sometimes called the *highest set* $\{H\}$. For a canonical weight vector, there exist smaller minimal and maximal sets of vertices, which are sometimes called the *canonically minimal* and *canonically maximal sets* of vertices, respectively.

It is of interest to note that the minimal set for a nonnegative weight vector is revealed by the irredundant normal disjunctive form of the function, and can be easily found. Similarly the maximal set is revealed by the irredundant normal disjunctive form of the complementary function $\bar{F}$.

THEOREM 5.4.1. Let a regular switching function F of n variables be expressed in an irredundant normal disjunctive form. Group the terms according to their length (i.e. number of literals in the simple products): G_1 contains those terms with one literal, G_2 contains those terms with two literals, etc. Consider a pair of variables x_j and x_k.

(1) F is symmetrical in x_j and x_k iff x_j and x_k appear equally often in each G.

(2) $c_j < c_k$ iff x_j and x_k appear unequally often in at least one group G_i and x_k appears more often than x_j in G_m, where m is the smallest index of such kind of groups.

PROOF:

(1) *Necessity* is obvious.

Sufficiency. Expanding F along x_j and x_k, we have

$$F = x_j x_k F_{x_j x_k} + x_j \bar{x}_k F_{x_j \bar{x}_k} + \bar{x}_j x_k F_{\bar{x}_j x_k} + \bar{x}_j \bar{x}_k F_{\bar{x}_j \bar{x}_k}$$

Since F is regular,

$$F = x_j x_k F_{x_j x_k} + x_j F_{x_j \bar{x}_k} + x_k F_{\bar{x}_j x_k} + F_{\bar{x}_j \bar{x}_k}$$

$x_j x_k F_{x_j x_k}$ and $F_{\bar{x}_j \bar{x}_k}$ are each symmetric in x_j and x_k, and x_j and x_k always appear equally often in $x_j x_k F_{x_j x_k}$ and also in $F_{\bar{x}_j \bar{x}_k}$.

$F_{x_j \bar{x}_k}$ and $F_{\bar{x}_j x_k}$ are comparable. Assume that $F_{x_j \bar{x}_k} \subset F_{\bar{x}_j x_k}$. Consider P, a term of length m in $F_{x_j \bar{x}_k}$. It either implies a term Q of length $m' \leq m$ in $F_{\bar{x}_j x_k}$, or implies the disjunction of several terms in $F_{\bar{x}_j x_k}$. In the latter case, by Theorem 4.2.4, the disjunction is not unate, and consequently F is also not unate, a contradiction to the fact that F is regular.

If P implies a single term Q with a length $m' = m$, it is obvious that $Q = P$.

Now consider the case that P implies a single Q with a length $m' < m$. Let us start with G_s, the group of terms with the shortest length s. Then $m = s-1$, and Q with a length of $m' < m$ does not exist. Hence for each P in $F_{x_j\bar{x}_k}$ there is also a $Q = P$ in $F_{\bar{x}_jx_k}$. Since x_j and x_k appear equally often in G_s,

$$(F_{x_j\bar{x}_k})_{s-1} = (F_{\bar{x}_jx_k})_{s-1}$$

where $(F_{x_j\bar{x}_k})_{s-1}$ and $(F_{\bar{x}_jx_k})_{s-1}$ are the disjunctions of terms with a length of $s-1$, in $F_{x_j\bar{x}_k}$ and $F_{\bar{x}_jx_k}$, respectively.

Next consider G_{s+1}. If a term P of length s in $F_{x_j\bar{x}_k}$ implies a term Q of length $s-1$ in $F_{\bar{x}_jx_k}$, there exists a Q in $F_{x_j\bar{x}_k}$ too, and x_jP implies x_jQ. Hence x_jP is not a prime implicant, a contradiction. Therefore Q must also be of the same length as P, and $Q = P$ again. Since x_j and x_k appear equally often in G_{s+1},

$$(F_{x_j\bar{x}_k})_s = (F_{\bar{x}_jx_k})_s$$

The same reasoning can be extended to G_{s+2}, G_{s+3}, ..., up to G_l, the group of the longest length. Therefore we have

$$F_{x_j\bar{x}_k} = F_{\bar{x}_jx_k}$$

If originally we assume that $F_{x_j\bar{x}_k} \supset F_{\bar{x}_jx_k}$, we shall arrive at the same conclusion. Hence F is symmetrical in x_j and x_k.

(2) *Necessity*. It is obvious that x_j and x_k must appear unequally often, for otherwise F is symmetrical in x_j and x_k, and $c_j = c_k$, a contradiction.

Since $c_j < c_k$, by Theorem 4.3.2,

$$F_{x_j\bar{x}_k} \subset F_{\bar{x}_jx_k}, \quad \text{and} \quad F_{x_j\bar{x}_k} \not\supset F_{\bar{x}_jx_k}$$

Consider $x_jF_{x_j\bar{x}_k}$ and $x_kF_{\bar{x}_jx_k}$. For those groups of terms from G_s through G_{m-1}, x_j and x_k appear equally often. By the same reasoning as in Part (1), those groups of $F_{x_j\bar{x}_k}$ and $F_{\bar{x}_jx_k}$ are equal, or

$$(F_{x_j\bar{x}_k})_{s,\ldots,m-1} = (F_{\bar{x}_jx_k})_{s,\ldots,m-1}$$

Next consider a term P of length $m-1$ in $F_{x_j\bar{x}_k}$, i.e. x_jP is a term in G_m. P cannot imply a disjunction of several terms in $F_{\bar{x}_jx_k}$, because otherwise F would not be unate. Hence P implies a single term Q of length $l \leqslant m-1$. If $l < m-1$, Q appears in the groups G_s through G_{m-1}, and is therefore also a term in $F_{x_j\bar{x}_k}$. Then x_jP implies x_jQ and is not a prime implicant, a contradiction. Hence $l = m-1$, and $Q = P$. Thus for every P of length $m-1$ in $F_{x_j\bar{x}_k}$, there exists the same term in $F_{\bar{x}_jx_k}$, which means that x_j cannot appear more often than x_k. Since in G_m x_j and x_k appear unequally often, x_k must appear more often than x_j.

Sufficiency. If x_k appears more often than x_j in G_m, there exists at least one term P of length $m-1$ in $F_{\bar{x}_j x_k}$ such that there does not exist a term $Q = P$ in $F_{x_j \bar{x}_k}$. By the same reasoning as above, P cannot imply a disjunction of several terms, and also cannot imply a term Q of length $l < m-1$ in $F_{x_j \bar{x}_k}$. Therefore

$$F_{x_j \bar{x}_k} \not\supset P, \quad \text{and} \quad F_{x_j \bar{x}_k} \not\supset F_{\bar{x}_j x_k}$$

But $F_{x_j \bar{x}_k}$ and $F_{\bar{x}_j x_k}$ are comparable. Therefore

$$F_{x_j \bar{x}_k} \subset F_{\bar{x}_j x_k}$$

By Theorem 4.3.2, $c_j < c_k$. Q.E.D.

Theorem 5.4.1 provides an easy means of determining the ordering of weights from the irredundant normal disjunctive form of a given function F, if F is known to be regular.

The concept of equivalence class and canonical form is helpful in that in testing and realization we need only consider functions in unnegated variables, all with positive weights, without loss of generality. However, the order of the variables is usually opposite to that of the canonical form, i.e. the subscripts of the variables and consequently of the weights are in ascending order as the descending order of the subscripts of the c's. This is quite natural because in writing an expression of a function or in expanding it along variables it is most convenient and reasonable to start from the variable with the highest value of c and to call it x_1. So later in testing and realization we shall assume that $a_1 \geq a_2 \geq \ldots \geq a_n \geq 0$.

5.5 COMPLEMENTARY, DUAL, AND COMPLEMENTARY-DUAL FUNCTIONS

THEOREM 5.5.1. If F is a threshold function of n variables, then the *complementary function* $\bar{F}$ is also a threshold function of n variables.

PROOF: Let $R[F] = (A;\ T)$ be a realization of F, and $T_l < T \leq T_u$, where T_l and T_u are the lower and upper boundaries of threshold value, respectively.

$$X_j A^t \geq T_u, \qquad X_j \in F$$

$$X_k A^t \leq T_l, \qquad X_k \in \bar{F}$$

Or

$$X_k(-A)^t \geq -T_l, \qquad X_k \in \bar{F}$$

$$X_j(-A)^t \leq -T_u, \qquad X_j \in F = (\bar{\bar{F}})$$

Since $T_u > T_l$, $-T_l > -T_u$, and there exists a threshold value T' such that $-T_u < T' \leq -T_l$. Therefore the above inequalities satisfy the definition of a threshold function, or $\bar{F}$ is a threshold function. Q.E.D.

COROLLARY 5.4.1–1. If $R[F] = (a_1, \ldots, a_n; T)$ is a realization of a threshold function F of n variables, where $T_l < T \leq T_u$, then $R[\bar{F}] = (a'_1, \ldots, a'_n; T')$ is a realization of the complementary function $\bar{F}$, where $a'_i = -a_i$, and $-T_u = T'_l < T' \leq T'_u = -T_l$.

COROLLARY 5.4.1–2. If $R[F] = (a_1, \ldots, a_n; T)$ is an integral realization of a threshold function F of n variables, where $T_l < T \leq T_u$, then $R[\bar{F}] = (-a_1, \ldots, -a_n; -T+1)$ is an integral realization of $\bar{F}$.

PROOF: For an integral realization, since $T_l < T \leq T_u$, $T_l < T-1 \leq T_u$. Thus we have

$$X_j A^t > T-1, \qquad X_j \in F$$

$$X_k A^t \leq T-1, \qquad X_k \in \bar{F}$$

Or

$$X_k(-A)^t \geq -T+1, \qquad X_k \in \bar{F}$$

$$X_j(-A)^t < -T+1, \qquad X_k \in F = \overline{(\bar{F})}$$

Hence $R[\bar{F}] = (-a_1, \ldots, -a_n; -T+1)$ is an integral realization of $\bar{F}$.

Q.E.D.

It can also be shown that the complementation of a switching function preserves linear separability, k-summability, k-asummability, k-monotonicity, and complete monotonicity. The preservation of linear separability has been shown above. The preservation of summability and asummability is obvious from the definitions of summability and asummability. The preservation of monotonicity is because of the fact that if G_1 and G_2 are any two reduced functions of a switching function F expanded along any subset of variables and if G'_1 and G'_2 are the corresponding reduced functions of $\bar{F}$, then $G_1 \supset G_2$ if $G'_1 \subset G'_2$, and $G_1 \subset G_2$ if $G'_1 \supset G'_2$. This latter fact is obvious from Lemma 4.4.2.

The *dual switching function* F^D of an arbitrary switching function F of n variables is defined as

$$F^D(x_1, \ldots, x_n) = \bar{F}(\bar{x}_1, \ldots, \bar{x}_n),$$

i.e. F^D is the complementary function $\bar{F}$ with all the variables complemented again.

Since the complementary function $\bar{F}$ can be obtained from any expression of F by the interchange of addition and multiplication operations and the complementation of the variables, the dual function can be obtained from F by simply interchanging the addition and multiplication operations.

The *complementary-dual switching function* $(\bar{F})^D$ of an arbitrary switching function F of n variables is defined as the dual function of the complementary function of F, i.e.

$$\bar{F}^D(x_1, \ldots, x_n) = \bar{\bar{F}}(\bar{x}_1, \ldots, \bar{x}_n)$$
$$= F(\bar{x}_1, \ldots, \bar{x}_n)$$

The *dual-complementary switching function* $(\overline{F^D})$ of an arbitrary switching function F of n variables is defined as the complementary function of the dual function of F, i.e.

$$(\overline{F^D})(x_1, \ldots, x_n) = \bar{\bar{F}}(\bar{x}_1, \ldots, \bar{x}_n)$$
$$= F(\bar{x}_1, \ldots, \bar{x}_n)$$

Thus it is seen that the dual-complementary function is the same as the complementary-dual function. It can be obtained from any expression of F by simply complementing all the variables. Therefore it belongs to the same equivalence class of F.

The operation of transforming a function into its dual function is called *dualization.* The operation of transforming a function into its complementary-dual function is called *complementation-dualization*, or *dualization-complementation.*

It is obvious that complementation-dualization, and consequently dualization, preserves linear separability, k-summability, k-asummability, k-monotonicity, and complete monotonicity.

THEOREM 5.5.2. If $R[F] = (a_1, \ldots, a_n;\ T)$, where $T_l < T \leq T_u$, is a realization of a threshold function F of n variables, then $R[\bar{F}^D] = (a_1', \ldots, a_n';\ T')$, where $a_i' = -a_i$, $T_l - \sum_{i=1}^{n} a_i = T_l' < T' \leq T_u' = T_u - \sum_{i=1}^{n} a_i$, is a realization of the complementary-dual function $\bar{F}$.

PROOF: This is a direct consequence of Theorem 5.1.1. Q.E.D.

THEOREM 5.5.3. If F is a threshold function of n variables, then the dual function F^D is also a threshold function of n variables.

PROOF: $\bar{F}$ is a threshold function, and F^D belongs to the same equivalence class of $\bar{F}$. Hence it is a threshold function.

THEOREM 5.5.4. If $R[F] = (a_1, \ldots, a_n;\ T)$ where $T_l < T \leq T_u$, is a realization of a threshold function F of n variables, then $R[F^D] = (a_1', \ldots, a_n';\ T')$, where $a_i' = a_i$ and $\sum_{i=1}^{n} a_i - T_u = T_l' < T' \leq T_u' = \sum_{i=1}^{n} a_i - T_l$, is a realization of the dual function F^D.

PROOF: Let $R[\bar{F}^D] = (a_1'', \ldots, a_n''; T'')$ where $T_l'' < T'' \le T_u''$, be a realization of the complementary dual function $\bar{F}^D$.

By Corollary 5.5.1–1,

$$a_i' = -a_i''$$
$$T_l' = -T_u''$$
$$T_u' = -T_l''$$

But by Theorem 5.5.2,

$$a_i'' = -a_i$$
$$T_l'' = T_l - \sum_{i=1}^{n} a_i$$
$$T_u'' = T_u - \sum_{i=1}^{n} a_i$$

Hence

$$a_i' = a_i$$
$$T_l' = \sum_{i=1}^{n} a_i - T_u$$
$$T_u' = \sum_{i=1}^{n} a_i - T_l$$

Q.E.D.

COROLLARY 5.5.4–1. If $R[F] = (a_1, \ldots, a_n; T)$, where $T_l < T \le T_u$, is an integral realization of a threshold function F of n variables, then $R[F^D] = (a_1', \ldots, a_n'; T')$, where $a_i' = a_i$, $T' = \sum_{i=1}^{n} a_i - T + 1$, is an integral realization of the dual function F^D.

PROOF: For an integral realization, since $T_l < T \le T_u$, $T_l \le T-1 < T_u$. Hence

$$T' = \sum_{i=1}^{n} a_i - T + 1 \le \sum_{i=1}^{n} a_i - T_l = T_u'$$

And
$$T' = \sum_{i=1}^{n} a_i - T + 1 > \sum_{i=1}^{n} a_i - T_u = T_l'$$

Q.E.D.

COROLLARY 5.5.4–2. The dual function F^D of a threshold function F of n variables is isobaric to F.

A switching function F is said to be *dual-comparable* iff $F \supset F^D$ or $F \subset F^D$.

COROLLARY 5.5.4–3. Every threshold function is dual-comparable.

A switching function F of n variables is said to be *self-dual* is $F = F^D$.

COROLLARY 5.5.4–4. A threshold function F of n variables is self-dual iff the realization of F is such that $T_u > \frac{1}{2}\sum_{i=1}^{n} a_i$ and $T_l < \frac{1}{2}\sum_{i=1}^{n} a_i$.

5.6 FINDING THE THRESHOLD FUNCTION FROM A GIVEN REALIZATION

The testing and realization of threshold functions will be studied in the next chapter. Here we shall study the reverse problem, i.e. to find the threshold function F when a realization of F, $R[F]$, is given. Although the testing and realization problem is a strenuous one, the reverse problem is relatively simple. If we list all the possible summations of weights that are greater than or equal to the threshold value T and represent each summation by a minterm such that when a weight a_i is present an uncomplemented variable x_i is used and when a weight a_i is not present a complemented variable $\bar{x}_i$ is used, then the disjunction of the minterms will be the required threshold function.

In fact it is not necessary to list all the summations of weights and all the minterms in the function. We can have a simpler and more systematic method of obtaining the function F.[3]

Assume that the function F is canonical and has an ordering of weights,

$$a_1 \geq a_2 \geq \dots \geq a_n.$$

We can represent the summations by a flow graph. First we list the nodes in columns. A node N_{ij} denotes the jth node in the ith column. Starting from node N_{11}, we draw a directed arc from node N_{11} to node N_{21}, and denote it by a_1. Then we draw a directed arc a_2 from node N_{21} to node N_{31}, and so on, with the successive arcs denoted by weights with successively higher subscripts, until an arc a_k from node N_{k1} to node $N_{k+1,1}$ is reached, such that $\sum_{i=1}^{k+1} a_i \geq T$ and $\sum_{i=1}^{k} a_i < T$, where node $N_{k+1,1}$ is called a terminal node. Then go back to the node immediately preceding $N_{k+1,1}$, viz. N_{k1}, and branch out, denoting the successive arcs by weights starting from a_{k+1}. After coming to a terminal node, we go back and branch out again, until no more branching can be made.

Then consider all the paths from N_{11} to the terminal nodes. With each weight a_i replaced by the corresponding variable x_i, we form the "gain", i.e. the product of the variables along the path, for each path. It is easily seen that each such product represents a prime implicant of F, and the sum of the products for all the paths is the irredundant normal disjunctive form of F.

Let us consider an example.

$$R[F] = (6, 5, 4, 3, 2, 1; 10)$$

The flow graph for finding F from $R[F]$ is shown in Fig. 3. It is easily seen from the flow graph that the threshold function is

$$F = x_1x_2 + x_1x_3 + x_1x_4x_5 + x_1x_4x_6 + x_2x_3x_4 + x_2x_3x_5 + x_2x_3x_6 + x_2x_4x_5 + x_3x_4x_5x_6$$

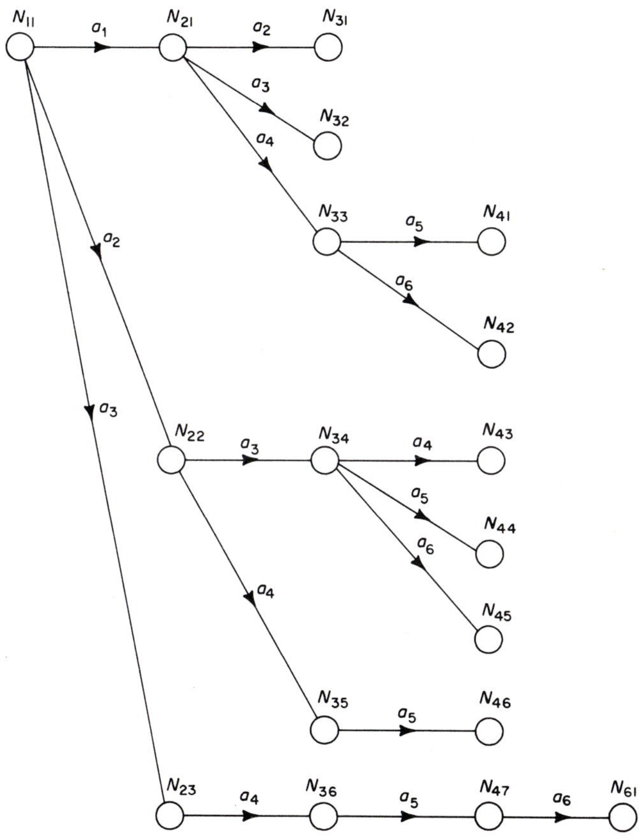

FIG. 3. Flow graph for finding a threshold function F from $R[F] = (6, 5, 4, 3, 2, 1; 10)$.

REFERENCES

1. Chow, C. K. (1961). On the characterization of threshold functions, in "Switching Circuit Theory and Logical Design", *A.I.E.E. Special Publication* **S134**, 34–38.
2. Elgot, C. C. (1961). Truth functions realizable by single threshold organs, in "Switching Circuit Theory and Logical Design", *A.I.E.E. Special Publication* **S134,** 225–245.
3. Fischler, M. A. (1962). Investigations concerning the theory and synthesis of linearly separable switching functions, Ph.D. dissertation, Department of Electrical Engineering, Stanford University, U.S.A.
4. Gabelman, I. J. (1962). The synthesis of Boolean functions using a single threshold element, *I.R.E. Trans. electronic Comput.* EC**11,** 639–642.
5. Gabelman, I. J. (1964). Properties and transformations of single threshold element functions, *I.E.E.E. Trans. electronic Comput.* EC**13,** 680–684.
6. Hu, S. T. (1965). "Threshold Logic", University of California Press, U.S.A.
7. McNaughton, R. (1961). Unate truth functions, *I.R.E. Trans. electronic Comput.* EC**10,** 1–6.
8. Muroga, S. (1961). Functional forms of majority decision functions and a necessary and sufficient condition for their realizability, in "Switching Circuit Theory and Logical Design", *A.I.E.E. Special Publication* **S134,** 39–46.
9. Muroga, S. (1965). Generation and asymmetry of self-dual threshold functions, *I.E.E.E. Trans. electronic Comput.* EC**14,** 125–136.
10. Muroga, S., Toda, I., and Takasu, S. (1961). Theory of majority decision elements, *J. Franklin Inst.* **271,** 376–418.
11. Slepian, D. (1953). On the number of symmetry types of Boolean functions of *n* variables, *Can. J. Math.* **5,** 185–193.
12. Winder, R. O. (1962). Threshold Logic, Doctoral dissertation, Mathematics Department, Princeton University, U.S.A.

CHAPTER 6

Testing and Realization of Threshold Functions

6.1 INTRODUCTION

In this chapter we shall study the testing and realization methods for threshold functions. Since a threshold function F of n variables is completely characterized by the Chow parameters $\{c_i\}$ and m, it was attempted to find a method of testing and realization from the properties of threshold functions, for such a method would be much simpler than the solution of a large set of inequalities, especially when n is large. Thus such properties as connectedness, unateness, regularity, monotonicity, etc. have been studied intensively. In fact it was shown by Winder[22] that complete monotonicity is a necessary and sufficient condition for a switching function of n variables to be a threshold function, for $n \leq 6$. However, complete monotonicity fails to be a sufficient condition when $n > 6$. Since linear separability is found to be equivalent to k-asummability for all values of k, it seems that there does not exist a necessary and sufficient condition that is not related to the inequalities. Therefore, except for trial and error methods, eventually we have to resort to the inequalities, no matter what preliminary testing steps applied to the function itself are taken.

The general methods for solving a set of inequalities have been summarized in detail by Hu.[11] In the following we shall give just a brief description of them. Since the testing and realization problem can be transformed into a minimization problem, we shall also explain briefly the simplex method of linear programming.[6, 11, 13, 14, 21] In addition to these rigorous mathematical methods, we shall describe a few practical methods. They include the tree expansion method by Coates and Lewis,[2, 3] the game theory method by Akers,[1] the trial method by Gabelman,[9] the approximation method by Dertouzos,[4, 5] and the map method by Torng.[20]

In the next chapter we shall present the successive higher ordering method, as originally developed by Sheng,[16] and later modified by Sheng and Hwa.[17]

Apart from these methods, the testing and realization of a threshold function can also be done by successive adjustments or approximations[11] of weights. This method is particularly suitable for pattern classification or adaptive systems, and will be discussed briefly in Chapter 10.

6.2 GENERAL METHODS

For a given function F of n variables, originally there are 2^n inequalities, corresponding to the 2^n vertices, out of which m inequalities correspond to vertices in F and $m' = 2^n - m$ inequalities correspond to vertices in $\bar{F}$.

Before the solution of the set of 2^n inequalities, we can reduce the number of inequalities to a minimum by eliminating the redundant inequalities.

If we know that the given function F is 1-monotonic, we can consider it to be positive in all the variables, without loss of generality. Then we can reduce the m inequalities to p inequalities, where $p \leq m$, corresponding to the set of minimal vertices in F, as defined for a positive 1-monotonic function, and similarly reduce the m' inequalities to q inequalities, where $q \leq m'$, corresponding to the set of maximal vertices in $\bar{F}$.†

For instance, for $n = 4$ and a nonnegative A, if $X = \bar{x}_1 x_2 x_3 x_4$ is a vertex of F, then $Y = x_1 x_2 x_3 x_4$ is also a vertex of F. While we keep the inequality $a_2 + a_3 + a_4 \geq T$ for X, the inequality $a_1 + a_2 + a_3 + a_4 \geq T$ for Y is redundant and may be deleted.

In addition, we have the following n constraining inequalities:

$$a_1 \geq 0$$
$$\cdots$$
$$a_n \geq 0$$

Therefore altogether there are $p + q + n$ inequalities instead of 2^n.

If the given function F is expressed in its irredundant normal disjunctive form, each term (or each prime implicant) P of F can be identified with a vertex by designating the kth position as 1 if x_k is present in P, and as 0 if it is not. This set of vertices obtained from the prime implicants is in fact the set of minimal vertices in F. Similarly, if the complementary function $\bar{F}$ is expressed in its irredundant normal disjunctive form, each term (or prime implicant) P of $\bar{F}$ can be identified with a vertex by designating the kth position as 0 if $\bar{x}_k$ is present, and as 1 if it is not. This set of vertices is in fact the set of maximal vertices in $\bar{F}$. Thus it is seen that the first round of reduction of the number of inequalities can be done directly on the functions F and $\bar{F}$, before the 2^n inequalities are written down, provided that F and $\bar{F}$ are expressed in their irredundant normal disjunctive forms. Alternately, of course, we may write down all the 2^n inequalities first and then delete the redundant ones.

If the given function F is known to be 2-monotonic, we can consider it to be canonical without loss of generality. Then we can further reduce the p inequalities to p' inequalities, where $p' \leq p$, corresponding to the set of canonically minimal vertices in F as defined for a regular function,

† The reasoning in this and the following several paragraphs is related to the notions and concepts introduced in Section 5.4, with which the reader is assumed to be familiar.

and reduce the q inequalities to q' inequalities, where $q' \leq q$, corresponding to the set of canonically maximal vertices in $\bar{F}$, as defined for a regular function.

For instance, for $n = 4$ and a canonical A, if $X = \bar{x}_1 x_2 x_3 x_4$ is a vertex of F, then $Y_1 = x_1 x_2 x_3 \bar{x}_4$, $Y_2 = x_1 x_2 \bar{x}_3 x_4$, $Y_3 = x_1 \bar{x}_2 x_3 x_4$ and $Y_4 = x_1 x_2 x_3 x_4$ are all vertices of F. While we keep the inequality $a_2 + a_3 + a_4 \geq T$ for X, the inequalities for Y_1, Y_2, Y_3, and Y_4, viz. $a_1 + a_2 + a_3 \geq T$, $a_1 + a_2 + a_4 \geq T$, $a_1 + a_3 + a_4 \geq T$, and $a_1 + a_2 + a_3 + a_4 \geq T$, are all redundant and may be deleted.

In addition, we have the following n constraining inequalities:

$$
\begin{aligned}
&a_1 - a_2 \geq 0 \\
&a_2 - a_3 \geq 0 \\
&\cdots \\
&a_{n-1} - a_n \geq 0 \\
&a_n \geq 0
\end{aligned}
$$

Therefore altogether there are $p' + q' + n$ inequalities instead of 2^n.

The solution of the set of inequalities is based on *Fan's principle of bounding solutions.*[8] The detailed explanation of the application of Fan's principle is given by Hu,[11] and will not be discussed here. We shall reproduce one of Hu's theorems as follows:

THEOREM 6.2.1. A switching function F of n variables is a threshold function iff in the set of inequalities there exist $n+1$ inequalities such that the $n+1$ linear equations obtained by replacing $>$ and $<$ by $=$ have a unique solution and that this unique solution satisfies all the inequalities.

THEOREM 6.2.1 is general in that it can be applied to the original set of 2^n inequalities, or to the reduced set of $p+q+n$ or $p'+q'+n$ inequalities.

It was proved in Hu[11] that if the given function is regular the equation $a_n = 0$ can be deleted unless $c_n = m/2$, i.e. F is independent of x_n, and the equation $a_i - a_{i+1} = 0$ can be deleted unless $c_i = c_{i+1}$, i.e. F is symmetrical in x_i and x_{i+1}. That this is true can also be easily seen from Theorem 5.3.5. Thus if F is not independent of x_n and is not symmetrical in any subset of variables, the n constraining inequalities are not used at all in the choice of $n+1$ linear equations, and consequently may be deleted. Thus the number of inequalities is further reduced to $p'+q'$.

In spite of its generality, the general method is not practicable unless n is very small, even for a reduced set of inequalities, because the family of all linearly independent sets of $n+1$ equations is very large, and the method depends on a search process through all the sets.

Another general method of solving the set of inequalities is by *successive elimination* of the unknowns T, a_n, ..., a_2. In the process of elimination certain inequalities will turn out to be redundant and can be deleted. Finally there will be a set of inequalities involving only one unknown a_1.

A value can be assigned to a_1 to satisfy this set of inequalities. By going backwards, values can be assigned to $a_2, \ldots, a_n$, and T successively.

The disadvantage of this method is the huge number of operations involved. The detailed procedure of this method is given in Hu[11] and will not be discussed here.

6.3 MINIMAL REALIZATION

A detailed exposition of minimal realization was given by Hu.[11] Here we shall discuss it briefly, giving only the most important conclusions without proof.

A realization for a threshold function F on n variables is said to be *normal* if $g = T_u - T_l = 1$.

Except for the realizations for the constant functions 0 and 1, the realization for any threshold function has a finite gap length g. Thus every realization can be normalized by dividing T and each a_i, $i = 1, \ldots, n$, by g.

To define minimal realization for a threshold function F of n variables, we specify a *cost function* K. Two most commonly used cost functions are

$$K_1 = \sum_{i=1}^{n} |a_i| + |T| \qquad \text{and} \qquad K_2 = \sum_{i=1}^{n} |a_i|$$

We shall denote the cost function for a realization $R[F]$ by $K(R)$.

A realization $R[F]$ is said to be *minimal* with respect to a cost function K iff $g \geq 1$ and for every other realization $R'[F]$ with $g \geq 1$, $K(R') \geq K(R)$.

In the following, when we say minimal realization, we mean minimal realization with respect to either of the above-mentioned cost functions, viz. K_1 or K_2.

It is obvious that a minimal realization for a threshold function F of n variables is necessarily normal, because otherwise K can be decreased by multiplying a factor of $1/g$ in normalization.

It is to be noted that with respect to a cost function K the minimal realization for a threshold function F of n variables is not necessarily unique. For certain functions there may exist more than one minimal realization.

It can be shown that for a canonical threshold function F of n variables there exists a *canonical minimal realization.*

A canonical realization for a threshold function F is said to be *strictly canonical* if $a_i = 0$ whenever F is independent of x_i and $a_j = a_k$ whenever F is symmetrical in x_j and x_k.

It can also be shown that for a canonical threshold function F of n variables there exists a *strictly canonical minimal realization.*

Winder[22] discovered an interesting example, a threshold function F

of 8 variables, which can have numerous minimal realizations. The strictly canonical minimal realization of F (with weights in descending order as the ascending order of subscripts) is

$$R[F] = (9, 6\tfrac{1}{2}, 6\tfrac{1}{2}, 5, 5, 3, 2, 2; 13)$$

It is seen that any realization with $a_1 = 9$, $6 \leq a_2 \leq 7$, $6 \leq a_3 \leq 7$, $a_2 + a_3 = 13$, $a_4 = 5$, $a_5 = 5$, $a_6 = 3$, $a_7 = 2$, $a_8 = 2$, and $T = 13$ is also minimal.

It is also of interest to note that complementation of variables does not always preserve the minimality of realization with respect to the cost function K_1. For instance, if the variable x_2 in the above example is complemented to obtain a function F', the realization becomes

$$R[F'] = (9, -6\tfrac{1}{2}, 6\tfrac{1}{2}, 5, 5, 3, 2, 2; 6\tfrac{1}{2})$$

which is not minimal with respect to K_1. However, the realization

$$R'[F'] = (9, -7, 6, 5, 5, 3, 2, 2; 6)$$

is minimal.

It can be seen that complementation of variables does preserve the minimality of realization with respect to the cost function K_2.

A minimal realization $R[F]$ for a threshold function F of n variables is said to be *uniquely minimal* with respect to a cost function K iff for every other realization $R'[F]$ with $g \geq 1$, $K(R') > K(R)$.

The above function obviously does not have a uniquely minimal realization.

Consider another threshold function F of 9 variables, which was discovered by Willis and quoted by Hu[11] with a realization

$$R[F] = (9, 6\tfrac{1}{2}, 6\tfrac{1}{2}, 6\tfrac{1}{2}, 5, 5, 3, 2, 2; 13)$$

It can be shown that this realization is uniquely minimal.

It is obvious that for any integral realization, $g \geq 1$, because the smallest integral gap length is 1.

An integral realization $R[F]$ for a threshold function F of n variables is said to be *minimal-integral* with respect to a cost function K iff for any other integral realization $R'[F]$, $K(R') \geq K(R)$.

It is to be noted that a minimal-integral realization is not necessarily minimal. For instance, for the above function F of 9 variables, the integral realization

$$R[F] = (9, 7, 7, 6, 5, 5, 3, 2, 2; 13)$$

is minimal-integral, but not minimal.

However, it can be shown that for a threshold function F of n variables which is not symmetric in any subset of variables, the minimal realization with respect to a cost function K is always integral, i.e. the minimal-integral realization is also minimal.

6.4 REALIZATION BY MINIMIZATION

Since every threshold function has a minimal realization, the realization problem can be solved by solving the minimization problem, i.e. by finding a minimal realization with respect to a given cost function.

Considering physical realization, an integral realization is preferred to a fractional one. Besides, for a threshold function not symmetric in any subset of variables, the minimal realization is also integral. Even for a threshold function symmetric in some subsets of variables, the minimal-integral realization is never too far from the minimal realization. Therefore in the following we shall consider integral realization only.

It can be shown that a regular switching function F of n variables is a threshold function iff it has a canonical minimal realization. In view of this fact, we can include the testing problem in the minimization problem. If we find a canonical minimal realization for F, F is certainly a threshold function. On the other hand, if we prove that F has no canonical minimal realization, then F is not a threshold function.

Thus the minimization problem becomes the problem of testing whether or not a given function F of n variables is a threshold function, and, if F is a threshold function, finding a minimal realization for F with respect to a given cost function K.

A standard problem in linear programming is to find nonnegative real numbers which minimize (maximize) a given linear function subject to a given set of linear inequalities. For a canonical 2-monotonic switching function F, the testing of linear separability and the minimal realization can be put in the form of such a standard minimizing problem.

The theory of the *simplex method of linear programming* and its application to threshold logic can be found in the literature.[6, 11, 13, 14] Here we shall give a brief explanation of this method applied to the testing and realization of threshold functions, without giving the rigorous justification.

For convenience, we shall denote the threshold value T by a_0.

We shall use K_1 as the cost function. Since the weights are all non negative, a_0 is also non negative, for otherwise the threshold function realized is the constant function 1. Thus we have

$$K_1 = \sum_{i=0}^{n} |a_i| = \sum_{i=0}^{n} a_i \tag{6.4–1}$$

Let

$\{X_j\}, j = 1, 2, \ldots, p'$ be the canonically minimal set of vertices of F.

$\{X_k\}, k = p'+1, \ldots, p'+q'$, be the canonically maximal set of vertices of $\bar{F}$.

$X'_j = (0, -1, x_{j1}, x_{j2}, \ldots, x_{jn})$ be a $(n+2)$-vector corresponding to a vertex X_j in F.

$X'_k = (1, -1, x_{k1}, x_{k2}, \ldots, x_{kn})$ be a $(n+2)$-vector corresponding to a vertex X_k in $\bar{F}$.

$A' = (1, a_0, a_1, \ldots, a_n)$ be a $(n+2)$-vector, where a_0 is the threshold value and a_i, $i = 1, \ldots, n$, is the weight of x_i.

$a_0, \ldots, a_n$ are the quantities to be determined from the inequalities. To distinguish from the binary variables x_i, they are called the unknowns of the inequalities.

Since $X_j A^t \geq a_0$, we have

$$X'_j A'^t \geq 0, \quad j = 1, \ldots, p' \tag{6.4–2}$$

Similarly, since $X_k A^t \leq a_0$, for integral realization,

$$X_k A^t \leq a_0 - 1.$$

Hence we have

$$X'_k A'^t \leq 0$$

Or

$$(-X'_k) A'^t \geq 0, \quad k = p'+1, \ldots, p'+q' \tag{6.4–3}$$

Since the weight vector is canonical, we have

$$a_1 - a_2 \geq 0$$

$$\cdots$$

$$a_{n-1} - a_n \geq 0$$

$$a_n \geq 0$$

Or

$$\begin{bmatrix} 0 & 0 & 1 & -1 & 0 & \ldots & 0 \\ 0 & 0 & 0 & 1 & -1 & 0 \ldots & 0 \\ \cdots & & & & & & \\ 0 & 0 \ldots & & & & & 1 \end{bmatrix} \begin{bmatrix} 1 \\ a_0 \\ a_1 \\ \vdots \\ a_n \end{bmatrix} \geq 0 \tag{6.4–4}$$

(6.4–2) consists of p' inequalities, (6.4–3) consists of q' inequalities, and (6.4–4) consists of n inequalities. Altogether there are $p'+q'+n$ inequalities. Putting them together in one matrix form, we have

$$\begin{bmatrix} 0 & -1 & x_{11} \cdots & & x_{1n} \\ \cdots & & & & \\ 0 & -1 & x_{p'1} \cdots & & x_{p'n} \\ -1 & 1 & -x_{p'+1,1} \cdots & & -x_{p'+1,n} \\ \cdots & & & & \\ -1 & 1 & -x_{p'+q',1} \cdots & & -x_{p'+q',n} \\ 0 & 0 & 1 \; -1 & & 0 \ldots 0 \\ \cdots & & & & \\ 0 & 0 \ldots & & & 1 \end{bmatrix} \begin{bmatrix} 1 \\ a_0 \\ a_1 \\ \vdots \\ a_n \end{bmatrix} \geq 0 \qquad (6.4\text{–}5)$$

Or

$$MA''^t \geq 0$$

where M is a $(p'+q'+n)\times(n+2)$-matrix.

Now the minimization problem is to find a "weight" vector A' such that inequalities (6.4–5) are all satisfied and the cost function K_1 in (6.4–1) is a minimum.

It can be shown that inequalities (6.4–4) are redundant with inequalities (6.4–2) and (6.4–3), and therefore may be deleted. Thus we have

$$M'A''^t \geq 0 \qquad (6.4\text{–}6)$$

where M' is a $(p'+q')\times(n+2)$-matrix.

Now set $y_m = X'_m A''^t$, $m = 1, \ldots, p'+q'$, where $\{y_m\}$ are called *slack variables.*

It is seen that if the elements of the first column of M' are all nonnegative, then the solution that $a_i = 0$, $i = 0, 1, \ldots, n$, will satisfy (6.4–6). This is called a *trial solution.* The simplex method is a sequence of trial solutions and substitutions of slack variables.

If a trial solution fails, we check all the equations $y_m = X'_m A''^t$ with a negative constant term on the right-hand side to see if there exists any equation with all the coefficients of a_i, $i = 0, \ldots, n$, nonpositive. If so, the minimization problem has no solution, because $a_i \geq 0$, $i = 0, \ldots, n$. If not, take any equation $y_m = X'_m A''^t$ with one or more positive coefficients. Then choose an unknown a_i and solve for a_i in terms of $a_0, a_1, \ldots,$

$a_{i-1}, a_{i+1}, \ldots, a_n$, and y_m. The choice of a_i is such that after substituting this expression for a_i into the cost function K_1, the resulting expression for K_1 will have a minimum increase in the constant term.

Next substitute the expression for a_i into inequalities (6.4–6), which become

$$M'_1 A_1'^{\,t} \geq 0$$

where $A'_1 = (1, a_0, a_1, \ldots, a_{i-1}, y_m, a_{i+1}, \ldots, a_n)$ and M'_1 is also different from M'.

Then repeat a trial solution that $a_0 = a_1 = \ldots = y_m = \ldots = a_n = 0$. If it still fails, then repeat the whole procedure again. In this way we obtain successive transformations of the matrices and a sequence of matrix inequalities

$$M'_1 A_1'^{\,t} \geq 0, M'_2 A_2'^{\,t} \geq 0, \ldots, M'_r A_r'^{\,t} \geq 0$$

Eventually either one of the two following possibilities will result: (1) The elements of the first column of M'_r are all nonnegative. Then by setting

$$A'_r = (1, 0, 0, \ldots, 0)$$

we can solve for A' which corresponds to the minimal-integral realization of F. (2) There exists a $y_m = X'_m A_r'^{T}$ with a negative constant term and all nonpositive coefficients on the right-hand side. In this case the minimization problem has no solution and consequently F is not a threshold function.

We shall illustrate this method by a simple example.

EXAMPLE

$$\begin{aligned} F &= x_1x_2 + x_1x_3 + x_2x_3x_4 \\ \bar{F} &= (\bar{x}_1 + \bar{x}_2)(\bar{x}_1 + \bar{x}_3)(\bar{x}_2 + \bar{x}_3 + \bar{x}_4) \\ &= \bar{x}_1\bar{x}_2 + \bar{x}_1\bar{x}_3 + \bar{x}_1\bar{x}_4 + \bar{x}_2\bar{x}_3 \end{aligned}$$

F is regular and is symmetric in x_2 and x_3. By Theorem 5.3.5,

$$a_1 > \begin{matrix} a_2 \\ a_3 \end{matrix} > a_4 > 0$$

Since for a minimal-integral realization, a_2 is not necessarily equal to a_3, we do not give ordering between a_2 and a_3, although it is admissible that $a_2 = a_3$. Thus in inequalities (6.4–4) we have $n+1$ instead of n inequalities.

The canonically minimal set of F and the canonically maximal set of $\bar{F}$ are, respectively,

$$\{X_j\} = \{1100, 1010, 0111\}$$

$$\{X_k\} = \{1001, 0110\}$$

$$K_1 = a_0 + a_1 + a_2 + a_3 + a_4 \geq 0$$

The inequalities, with the left-hand side equated to the slack variables, are

$$\begin{aligned}
y_1 &= \quad -a_0 + a_1 + a_2 && \geq 0 \\
y_2 &= \quad -a_0 + a_1 \quad + a_3 && \geq 0 \\
y_3 &= \quad -a_0 \quad + a_2 + a_3 + a_4 && \geq 0 \\
y_4 &= -1 + a_0 - a_1 \quad - a_4 && \geq 0 \\
y_5 &= -1 + a_0 \quad - a_2 - a_3 && \geq 0 \\
y_6 &= \quad + a_1 - a_2 && \geq 0 \\
y_7 &= \quad + a_1 \quad - a_3 && \geq 0 \\
y_8 &= \quad + a_2 \quad - a_4 && \geq 0 \\
y_9 &= \quad + a_3 - a_4 && \geq 0 \\
y_{10} &= \quad + a_4 && \geq 0
\end{aligned}$$

It is also admissible to assume that $a_2 \geq a_3$. In that case there would be 9 instead of 10 inequalities. However, this is immaterial because among the above 10 inequalities the last 5 are redundant. It is easily seen that y_6 is redundant with y_2 and y_5, y_7 is redundant with y_1 and y_5, y_8 is redundant with y_1 and y_4, y_9 is redundant with y_2 and y_4, and y_{10} is redundant with y_3 and y_5. After deleting the last 5 inequalities, we have

$$M' = \begin{array}{c} \\ y_1 \\ y_2 \\ y_3 \\ y_4 \\ y_5 \end{array} \begin{array}{c} \begin{array}{cccccc} \text{Constant} & a_0 & a_1 & a_2 & a_3 & a_4 \end{array} \\ \left[\begin{array}{cccccc} 0 & -1 & 1 & 1 & 0 & 0 \\ 0 & -1 & 1 & 0 & 1 & 0 \\ 0 & -1 & 0 & 1 & 1 & 1 \\ -1 & \textcircled{1} & -1 & 0 & 0 & -1 \\ -1 & 1 & 0 & -1 & -1 & 0 \end{array}\right] \end{array}$$

$$A' = (1, a_0, a_1, a_2, a_3, a_4)$$

In matrix M', we denote the rows by the slack variables $\{y_m\}$, and denote the columns by "constant" and the unknowns $\{a_i\}$.

Row y_4 has a negative element in the first column and a positive coefficient for a_0, which is circled. Solving for a_0, we have

$$a_0 = y_4 + 1 + a_1 + a_4$$

After substitution, we have

$$M_1' = \begin{array}{c} \\ y_1 \\ y_2 \\ y_3 \\ y_4 \\ y_5 \end{array} \begin{array}{c} \begin{array}{cccccc} \text{Constant} & y_4 & a_1 & a_2 & a_3 & a_4 \end{array} \\ \begin{bmatrix} -1 & -1 & 0 & \textcircled{1} & 0 & -1 \\ -1 & -1 & 0 & 0 & 1 & -1 \\ -1 & -1 & -1 & 1 & 1 & 0 \\ 0 & 1 & 0 & 0 & 0 & 0 \\ 0 & 1 & 1 & -1 & -1 & 1 \end{bmatrix} \end{array}$$

$A_1' = (1, y_4, a_1, a_2, a_3, a_4)$

Row y_1 has a negative element in the first column and a positive coefficient for a_2, which is circled. Solving for a_2, we have

$$a_2 = y_1+1+y_4+a_4$$

After substitution, we have

$$M_2' = \begin{array}{c} \\ y_1 \\ y_2 \\ y_3 \\ y_4 \\ y_5 \end{array} \begin{array}{c} \begin{array}{cccccc} \text{Constant} & y_4 & a_1 & y_1 & a_3 & a_4 \end{array} \\ \begin{bmatrix} 0 & 0 & 0 & 1 & 0 & 0 \\ -1 & -1 & 0 & 0 & \textcircled{1} & -1 \\ 0 & 0 & -1 & 1 & 1 & 1 \\ 0 & 1 & 0 & 0 & 0 & 0 \\ -1 & 0 & 1 & -1 & -1 & 0 \end{bmatrix} \end{array}$$

$A_2' = (1, y_4, a_1, y_1, a_3, a_4)$

Row y_2 has a negative element in the first column and a positive coefficient for a_3, which is circled. Solving for a_3, we have

$$a_3 = y_2+1+y_4+a_4$$

After substitution, we have

$$M_3' = \begin{array}{c} \\ y_1 \\ y_2 \\ y_3 \\ y_4 \\ y_5 \end{array} \begin{array}{c} \begin{array}{cccccc} \text{Constant} & y_4 & a_1 & y_1 & y_2 & a_4 \end{array} \\ \begin{bmatrix} 0 & 0 & 0 & 1 & 0 & 0 \\ 0 & 0 & 0 & 0 & 1 & 0 \\ 1 & 1 & -1 & 1 & 1 & 2 \\ 0 & 1 & 0 & 0 & 0 & 0 \\ -2 & -1 & \textcircled{1} & -1 & -1 & -1 \end{bmatrix} \end{array}$$

$A_3' = (1, y_4, a_1, y_1, y_2, a_4)$

Row y_5 has a negative element in the first column and a positive coefficient for a_1, which is circled. Solving for a_1, we have

$$a_1 = y_5+2+y_4+y_1+y_2+a_4$$

After substitution, we have

$$M'_4 = \begin{array}{c} \\ y_1 \\ y_2 \\ y_3 \\ y_4 \\ y_5 \end{array} \begin{array}{c} \begin{array}{cccccc} \text{Constant} & y_4 & y_5 & y_1 & y_2 & a_4 \end{array} \\ \left[\begin{array}{cccccc} 0 & 0 & 0 & 1 & 0 & 0 \\ 0 & 0 & 0 & 0 & 1 & 0 \\ -1 & 0 & -1 & 0 & 0 & \textcircled{1} \\ 0 & 1 & 0 & 0 & 0 & 0 \\ 0 & 0 & 1 & 0 & 0 & 0 \end{array}\right] \end{array}$$

$A'_4 = (1, y_4, y_5, y_1, y_2, a_4)$

Row y_3 has a negative element in the first column and a positive coefficient for a_4, which is circled. Solving for a_4, we have

$$a_4 = y_3 + 1 + y_5$$

After substitution, we have

$$M'_5 = \begin{array}{c} \\ y_1 \\ y_2 \\ y_3 \\ y_4 \\ y_5 \end{array} \begin{array}{c} \begin{array}{cccccc} \text{Constant} & y_4 & y_5 & y_1 & y_2 & y_3 \end{array} \\ \left[\begin{array}{cccccc} 0 & 0 & 0 & 1 & 0 & 0 \\ 0 & 0 & 0 & 0 & 1 & 0 \\ 0 & 0 & 0 & 0 & 0 & 1 \\ 0 & 1 & 0 & 0 & 0 & 0 \\ 0 & 0 & 1 & 0 & 0 & 0 \end{array}\right] \end{array}$$

$A'_5 = (1, y_4, y_5, y_1, y_2, y_3)$

Now the first column is all nonnegative. By setting $y_4 = y_5 = y_1 = y_2 = y_3 = 0$, all the inequalities are satisfied. Solving for the weights, we have

$$a_4 = 1, a_1 = 3, a_3 = 2, a_2 = 2, a_0 = 5.$$

Therefore

$$K_1 = 5+3+2+2+1 = 13,$$

and

$$R[F] = (3, 2, 2, 1; 5)$$

6.5 TREE EXPANSION METHOD

Coates and Lewis[2,3] developed a *tree method* for testing and realization of threshold functions by expanding the given function F of n variables along the variables $x_1, \ldots, x_{n-1}$ successively in the form of a tree, thus obtaining a set of reduced functions of F. Here we shall explain briefly the basic principle of the method.

Assume that the given function F of n variables is a canonical 2-monotonic function. Expanding F along x_1, we have

$$\begin{aligned} F &= x_1 F_{x_1} + \bar{x}_1 F_{\bar{x}_1} \\ &= x_1 F_{x_1} + F_{\bar{x}_1} \end{aligned}$$

If F is a threshold function, then, by Theorem 4.3.4, F_{x_1} and $F_{\bar{x}_1}$ are isobaric functions of $n-1$ variables. Furthermore, if

$$R[F] = (a_1, a_2, \ldots, a_n; T),$$

then

$$R[F_{x_1}] = (a_2, \ldots, a_n; T')$$

$$R[F_{\bar{x}_1}] = (a_2, \ldots, a_n; T'')$$

where $T' = T - a_1$, and $T'' = T$.

If T' is increased by an amount a_1, it will be the same as T''. Since $T_l' < T' \leq T_u'$ and $T_l'' < T'' \leq T_u''$, if T_l' and T_u' are each increased by an amount a_1, then the gap length $g' = (T_u' + a_1) - (T_l' + a_1) = T_u' - T_l'$ remains the same, but the gap $T_u' + a_1 : T_l' + a_1$ have been raised by an amount a_1 and therefore will overlap or have a part common to the gap $T_u'' : T_l''$. Then there exists a threshold value $T = T'$ lying in this common part of gap.

Now the realization problem is the reverse one. If F_{x_1} and $F_{\bar{x}_1}$ are not comparable, or if F_{x_1} and $F_{\bar{x}_1}$ are comparable but are not isobaric to each other, then F is not a threshold function. If F_{x_1} and $F_{\bar{x}_1}$ are isobaric threshold functions and have a common weight vector $(a_2, \ldots, a_n)$, then T_u', T_l', T_u'', and T_l'' can be determined. If the gap $T_u' : T_l'$ can be raised by an amount a_1 such that the gap $T_u' + a_1 : T_l' + a_1$ and the gap $T_u'' : T_l''$ have a common part, then $F = x_1 F_{x_1} + F_{\bar{x}_1}$ is also a threshold function, and a_1 can be assigned as the weight of x_1.

In the tree expansion, the original function F is said to be at the 1st level, the functions F_{x_1} and $F_{\bar{x}_1}$ are said to be at the 2nd level, and the function $F_{x_1^* x_2^* \ldots x_k^*}$ is said to be at the $(k+1)^{\text{th}}$ level. It is obvious that if at a certain level there is only one pair of reduced functions F_{x_i} and $F_{\bar{x}_i}$ which are isobaric to each other, then there always exists an a_i and therefore $x_i F_{x_i} + F_{\bar{x}_i}$ is necessarily a threshold function. However, if at a certain level there is more than one pair of reduced functions, then a_i should be the same for every pair in order that all the reduced functions at the same level are all isobaric threshold functions. In this case a_i does not necessarily exist.

Considering one pair of reduced functions F_{x_1} and $F_{\bar{x}_1}$, if the gap $T_u' : T_l'$ is raised by an amount $T_l'' - T_u'$, then the two gaps will be just adjacent to each other. To obtain a certain amount of overlapping, gap $T_u' : T_l'$ should be raised by an amount $T_l'' - T_u' + g$, or $a_1 = T_l'' - T_u' + g$, where g may be taken to be $\text{Min}\{g', g''\}$ when either g' or g'' or both are finite. For integral realization, g can be taken to be ≥ 1. Thus

$$a_1 \leq T_l'' - T_u' + 1$$

It is to be noted that for a constant function 1, $T_u = 0$ and T_l is taken to be $-\infty$. For a constant function 0, $T_l = \sum_{i=1}^{n} a_i$ and T_u is taken to be ∞.

The realization procedure starts at the bottom, or the nth level, where reduced functions are functions of x_n only. If F is not independent of x_n, $a_n > 0$. For integral realization, as a first choice, we can assign $a_n = 1$.

It is to be noted that if a weight vector $A = (a_1, \ldots, a_n)$ can be found such that the requirement of gap overlapping is satisfied at every level, then F is a threshold function. But if at some level a weight a_k cannot be found, it does not necessarily imply that F is not a threshold function, because the previously determined weights $a_n, \ldots, a_{k+1}$ were not uniquely determined, but were chosen more or less by trial and error. Therefore, in that case, we have to repeat the procedure by making other trials.

We shall illustrate this method by an example.

EXAMPLE

$$F = x_1x_2 + x_1x_3 + x_2x_3x_4$$

The tree expansion of F along the variables x_1, x_2 and x_3 is shown below.

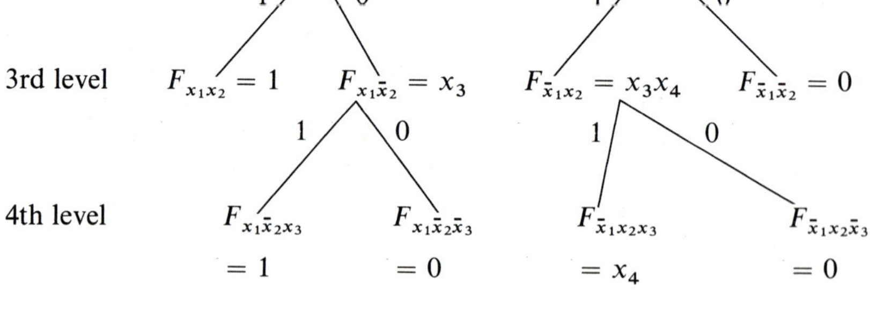

It is to be noted that when a reduced function is 1, all its reduced functions expanded along the remaining variables are 1, and therefore it is not necessary to continue the expansion any further. Similarly, when a reduced function is 0, all its reduced functions expanded along the remaining variables are 0.

Since at the 4th level, $F_{\bar{x}_1 x_2 x_3} = x_4$, F is not independent of x_4. So we assign $a_4 = 1$.

At the 4th level, there are two pairs of functions $\{F_{x_1\bar{x}_2x_3}, F_{x_1\bar{x}_2\bar{x}_3}\}$ and $\{F_{\bar{x}_1x_2x_3}, F_{\bar{x}_1x_2\bar{x}_3}\}$ to be considered.

For the first pair,

$$F_{x_1\bar{x}_2x_3} = 1, \qquad T_u' = 0, \qquad T_l' = -\infty$$

$$F_{x_1\bar{x}_2\bar{x}_3} = 0, \qquad T_u'' = \infty, \qquad T_l'' = a_4 = 1$$

We shall consider integral realization. Hence

$$a_3 \geq T_l'' - T_u' + 1 = 1 - 0 + 1 = 2$$

For the second pair,

$$F_{\bar{x}_1x_2x_3} = x_4, \qquad T_u' = 1, \qquad T_l' = 0$$

$$F_{\bar{x}_1x_2\bar{x}_3} = 0, \qquad T_u'' = \infty, \qquad T_l'' = 1$$

$$a_3 \geq T_l'' - T_u' + 1 = 1 - 1 + 1 = 1$$

So we assign $a_3 = 2$.

At the 3rd level, again there are two pairs of reduced functions to be considered, *viz.* $\{F_{x_1x_2}, F_{x_1\bar{x}_2}\}$ and $\{F_{\bar{x}_1x_2}, F_{\bar{x}_1\bar{x}_2}\}$.

For the first pair,

$$F_{x_1x_2} = 1, \qquad T_u' = 0, \qquad T_l' = -\infty$$

$$F_{x_1\bar{x}_2} = x_3, \qquad T_u'' = 2, \qquad T_l'' = 1$$

$$a_2 \geq T_l'' - T_u' + 1 = 1 - 0 + 1 = 2$$

For the second pair,

$$F_{\bar{x}_1x_2} = x_3x_4, \qquad T_u' = 3, \qquad T_l' = 2$$

$$F_{\bar{x}_1x_2} = 0, \qquad T_u'' = \infty, \qquad T_l'' = 3$$

$$a_2 \geq T_l'' - T_u' + 1 = 3 - 3 + 1 = 1$$

So we assign $a_2 = 2$.

At the 2nd level, there is only one pair of reduced functions to be considered, *viz.* $\{F_{x_1}, F_{\bar{x}_1}\}$.

$$F_{x_1} = x_2+x_3, \qquad T'_u = 2, \qquad T'_l = 1$$

$$F_{\bar{x}_1} = x_2x_3x_4, \qquad T''_u = 5, \qquad T''_l = 4$$

$$a_1 \geq T''_l - T'_u + 1 = 4-2+1 = 3$$

So we assign $a_1 = 3$, and $T = T''_u = 5$. Thus we have

$$R[F] = (3, 2, 2, 1; 5)$$

6.6 GAME-THEORY METHOD

Akers[1] developed a method for testing and realization of threshold functions using the *two-person zero-sum game theory*. Here we shall explain briefly the basic principle of this method, rather than giving the details of the procedure.

In a non-strictly determined game, $[G]$ is an $m \times n$ matrix. The number v is called its value. P, a $1 \times m$ row vector, and Q, an $n \times 1$ column vector, are optional strategies for the two players, respectively, if the following inequalities are satisfied:

$$PG = [p_1, p_2, \ldots, p_m][G] \leq (v, v, \ldots, v)$$

$$GQ = [G]\begin{bmatrix} q_1 \\ q_2 \\ \vdots \\ q_n \end{bmatrix} \geq \begin{bmatrix} v \\ v \\ \vdots \\ v \end{bmatrix}$$

In fact P and Q are probability distributions, and therefore are subject to the following constraints:

$$\sum_{i=1}^{m} p_i = 1$$

$$\sum_{i=1}^{m} q_i = 1$$

The game problem is a linear programming problem. If the game is a *non-strictly determined* one, there must exist a solution or a value v.

Akers put the testing and realization problem in the form of a game problem, with the weights normalized and the threshold value adjusted to 1/2. There is, however, a slight difference between these two kinds of problems. The testing and realization problem does not necessarily have a solution because the given function F is not necessarily a threshold function. For simplicity and to explain it in terms of integral realization, we shall relax the restrictions of Akers and present the problem in a modified form.

Considering the set of p' minimal vertices in F and the set of q' maximal vertices in $\overline{F}$, we have

$$
\begin{aligned}
X_j A^t &= \sum_{i=1}^{n} a_i x_{ji} \geq T_u, \qquad j = 1, \ldots, p' \\
X_k A^t &= \sum_{i=1}^{n} a_i x_{ki} \leq T_l, \qquad k = p'+1, \ldots, p'+q'
\end{aligned}
\tag{6.6–1}
$$

By setting $a_0 = T = (T_u + T_l)/2$, we have

$$T_u = a_0 + \tfrac{1}{2} g \qquad T_l = a_0 = \tfrac{1}{2} g$$

For minimal-integral realization, $g = 1$. Or

$$T_u = a_0 + \tfrac{1}{2} \qquad T_l = a_0 - \tfrac{1}{2}$$

Thus inequalities (6.6–1) become

$$
\begin{aligned}
X'_j A'^t &= \sum_{i=1}^{n} a_i x_{ji} \geq \tfrac{1}{2}, \qquad j = 1, \ldots, p' \\
(-X'_k) A'^t &= \sum_{i=0}^{n} a_i x_{ki} \geq \tfrac{1}{2}, \qquad k = p'+1, \ldots, p'+q'
\end{aligned}
\tag{6.6–2}
$$

where $X'_j = (-1, x_{j1}, \ldots, x_{jn})$ is a $1 \times (n+1)$ row vector.
$-X'_k = (1, -x_{k1}, \ldots, -x_{kn})$ is a $1 \times (n+1)$ row vector.
$A' = (a_0, a_1, \ldots, a_n)$ is a $1 \times (n+1)$ row vector.

Or

$$M' A'^t \geq \tfrac{1}{2} \tag{6.6–2′}$$

where M' is a $(p'+q') \times (n+1)$ matrix.

By setting $A''^t = A'^t / (\sum_{i=0}^{n} a_i)$, we have

$$M' A''^t \geq \frac{1}{2 \sum_{i=0}^{n} a_i} \tag{6.6–3}$$

Comparing with the game problem, we can see that M' corresponds to G, A''^t corresponds to Q, and $\dfrac{1}{2 \sum_{i=0}^{n} a_i}$ corresponds to v. We can define another $1 \times (p'+q')$ row vector B which is also to be determined such that

$$BM' \leq \frac{1}{2 \sum_{i=0}^{n} a_i}$$

Then by successive adjustments of A'''^t and B we can eventually arrive at a solution if it ever exists. However, if trial and error is permitted, it is not really necessary to have this row vector B.

Let $a_0' = T_u$

$X_j'' = (0, -1, x_{j1}, \ldots, x_{jn})$ be a $1 \times (n+2)$ row vector.

$-X_k'' = (-1, 1, -x_{k1}, \ldots, -x_{kn})$ be a $1 \times (n+2)$ row vector.

$A''' = (1, a_0', a_1, \ldots, a_n)$ be a $1 \times (n+2)$ row vector.

Then inequalities (6.6–1) become

$$\begin{aligned} X_j'' A'''^t &\geq 0, & j &= 1, \ldots, p' \\ (-X_k'') A'''^t &\geq 0, & k &= p'+1, \ldots, p'+q' \end{aligned} \qquad (6.6\text{–}4)$$

Or

$$M'' A'''^t \geq 0 \qquad (6.6\text{–}4')$$

where M'' is a $(p'+q') \times (n+2)$ matrix.

Since all the components of A'''^t are positive integers, the problem now reduces to that of trying to add up exactly one time of column 1 of M'' and any positive integral number of times of each columns 2, 3, ..., $n+2$, such that the sum of each row will be ≥ 0. If such positive integers exist, then the function F is a threshold function, and the set of integers are the threshold value and the weights of the variables. If such a set of positive integers does not exist, then the function F is not a threshold function.

We shall work out our previous Example again by using this method.

EXAMPLE

$$F = x_1 x_2 + x_1 x_3 + x_2 x_3 x_4$$

Using the set of minimal vertices in F and the set of maximal vertices in $\bar{F}$, we have

$$M'' A'''^t = \begin{array}{c} \begin{matrix} C_1 & C_2 & C_3 & C_4 & C_5 & C_6 \end{matrix} \\ \begin{bmatrix} 0 & -1 & 1 & 1 & 0 & 0 \\ 0 & -1 & 1 & 0 & 1 & 0 \\ 0 & -1 & 0 & 1 & 1 & 1 \\ -1 & 1 & -1 & 0 & 0 & -1 \\ -1 & 1 & 0 & -1 & -1 & 0 \end{bmatrix} \end{array} \begin{bmatrix} 1 \\ a_0 \\ a_1 \\ a_2 \\ a_3 \\ a_4 \end{bmatrix}$$

where the columns are denoted by $C_1, \ldots, C_6$.

By trial and error, it is found that

$$C_1+5C_2+3C_3+2C_4+2C_5+C_6 = 0$$

Now that $M''A'''^t = 0$, the condition is satisfied. So the function F is a threshold function. Since $A''' = (1, 5, 3, 2, 2, 1)$, we have

$$R[F] = (3, 2, 2, 1; 5)$$

6.7 A TRIAL METHOD

Gabelman[9] developed a trial method which is quite similar to the game-theory method. This method uses a succession of directed trials of values for the $\{a_i\}$, with the first trial set of values being the minimum set of $\{a_i\}$ allowed by the ordering of weights.

To begin with, we have $p'+q'$ inequalities corresponding to the set of minimal vertices in F and the set of maximal vertices in $\bar{F}$. Denote the left-hand side of the p' and q' inequalities, respectively, by

$$I_j = X_jA^t = \sum_{i=1}^{n} x_{ji}a_i, \qquad j = 1, \ldots, p'$$

and

$$I_k = X_kA^t = \sum_{i=1}^{n} x_{ki}a_i, \qquad k = p'+1, \ldots, p'+q'$$

where x_{ji} is the value of x_i in the vertex X_j.

x_{ki} is the value of x_i in the vertex x_k.

Construct a Table III, where the ith column is M_i, $i = 1, \ldots, n$; the jth row I_j; the kth row I_k; and wherein the (M_i, I_j) entry is $\sum_{h=1}^{i} x_{jh}$; the (M_i, I_k) entry is $\sum_{h=1}^{i} x_{kh}$.

The $(n+1)$th column is S_1, the (S_1, I_j) entry is $S_{j1} = \sum_{i=1}^{n}\left[\sum_{h=1}^{i} x_{jh}\right]$; the (S_1, I_k) entry is $S_{k1} = \sum_{i=1}^{n}\left[\sum_{h=1}^{i} x_{kh}\right]$.

$S_{j1}(S_{k1})$ is the value of $I_j(I_k)$ when the a_i are given the minimum values allowed by the ordering of weights. The $(n+d)$th column S_d has entries $S_{jd}(S_{kd})$, which are the values for $I_j(I_k)$ obtained as a result of the dth trial of values for a_i.

TABLE III

Trial Solution Format

I	M_1	M_2	...	M_n	S_1	...	S_d
I_1	x_{11}	$\sum_{h=1}^{2} x_{1h}$	...	$\sum_{h=1}^{n} x_{1h}$	S_{11}	...	S_{1d}
⋮	⋮	⋮					
$I_{p'}$	$x_{p',1}$	$\sum_{h=1}^{2} x_{p',h}$	...	$\sum_{h=1}^{n} x_{p',h}$	$S_{p',1}$	...	$S_{p',d}$
$I_{p'+1}$	$x_{p'+1,1}$	$\sum_{h=1}^{2} x_{p'+1,h}$	...	$\sum_{h=1}^{n} x_{p'+q',h}$	$S_{p'+1,1}$	...	$S_{p'+1,d}$
⋮	⋮	⋮		⋮	⋮		⋮
$I_{p'+q'}$	$x_{p'+q',1}$	$\sum_{h=1}^{2} x_{p'+q',h}$	...	$\sum_{h=1}^{n} x_{p'+q',h}$	$S_{p'+q',h}$	...	$S_{p'+q',d}$

A solution to the set of $p'+q'$ inequalities is obtained on the dth trial if each of the S_{jd} is greater than any of the S_{kd}.

The minimum integral values assignable as a first trial for the $\{a_i\}$ are $a_i^1 = (n-i+1)$. For this assignment, $S_{j1}(S_{k1})$ is the value of $I_j(I_k)$ since

$$S_{j1} = \sum_{i=1}^{n} \left[\sum_{h=1}^{i} x_{jh} \right]$$

$$= x_{j1} + (x_{j1} + x_{j2}) + \ldots + (x_{j1} + \ldots + x_{jn})$$
$$= nx_{j1} + (n-1)x_{j2} + \ldots + 2x_j(n-1) + x_{jn}$$
$$= \sum_{i=1}^{n} a_i^1 x_{ji}$$

A second trial assignment of values for the $\{a_i\}$, $\{a_i^2\}$, is obtained by choosing some m, adding each

$$\sum_{i=1}^{m} x_{ji} \quad (1 \leq m \leq n)$$

in the M_m column to the corresponding $S_{j1}(S_{k1})$ in S_2. Then

$$S_{j2} = S_{j1} + \sum_{i=1}^{m} x_{ji}$$
$$= \sum_{i=1}^{n} (a_i^1) x_{ji} + \sum_{i=1}^{m} x_{ji}$$
$$= \sum_{i=1}^{m} (a_i^1 + 1) x_{ji} + \sum_{i=m+1}^{n} a_i^1 x_{ji}$$

Thus in S_2, the $\{a_i\}$ have the values $\{a_i^2\}$ related to $\{a_i^1\}$ as follows:

$$a_i^2 = a_i^1 \qquad (i > m)$$
$$a_i^2 = a_i^1 + 1 \qquad (i \le m)$$

Or, in general,

$$a_i^{d+1} = a_i^d \qquad (i > m)$$
$$a_i^{d+1} = a_i^d + 1 \qquad (i \le m)$$

The procedure is thus to make successive trials until each of the S_{jd} is greater than any of the S_{kd}. The threshold value T is then set equal to the minimum S_{jd}.

If the column M_i is used t_i times in the synthesis, including original uses in forming S_1, then the weights $\{a_i\}$ are given by

$$a_i = \sum_{m=i}^{n} t_m$$

It is seen that this trial method is similar to the game-theory method in that the weights are monotonically increased in successive adjustments. This method is, however, a refinement over the game-theory method in that the ordering of weights, and consequently the differences between weights, are implicitly utilized. The concept of differences between weights will be fully explored in the successively higher ordering method, as will be discussed in Chapter 7.

We shall illustrate this method by the same example as above again.

EXAMPLE

$$F = x_1x_2 + x_1x_3 + x_2x_3x_4$$
$$\bar{F} = \bar{x}_1\bar{x}_2 + \bar{x}_1\bar{x}_3 + \bar{x}_1\bar{x}_4 + \bar{x}_2\bar{x}_3$$

F is symmetric in x_2 and x_3. However, it is not necessary to have $a_2 = a_3$, and we may assume that $a_2 > a_3$. Besides, since F is regular, we have

$$a_1 > a_2 > a_3 > a_4$$

Thus

$$\{X_j\} = \{X_1, X_2, X_3\} = \{1100, 1010, 0111\}$$
$$\{X_k\} = \{X_4, X_5\} = \{0110, 1001\}$$

The trial solution format is shown in Table IV, where the first column is added to indicate the sets of vertices $\{X_j\}, j = 1, 2, 3$, and $\{X_k\}, k = 4, 5$.

TABLE IV

X_j or X_k	I_j or I_k	M_1	M_2	M_3	M_4	S_1
$X_1 = 1100$	I_1	1	2	2	2	7
$X_2 = 1010$	I_2	1	1	2	2	6
$X_3 = 0111$	I_3	0	1	2	3	6
$X_4 = 0110$	I_4	0	1	2	2	5
$X_5 = 1001$	I_5	1	1	1	2	5

From the above table it is seen that each S_{j1} is greater than any of the S_{k1}. Therefore a solution is obtained on the first trial.

Thus we have

$$a_1 = 4, \quad a_2 = 3, \quad a_3 = 2, \quad a_4 = 1, \quad T = 6.$$

Or

$$R[F] = (4, 3, 2, 1; 6)$$

6.8 AN APPROXIMATION METHOD

As discussed in Chapter 3, Dertouzos[4, 5] has derived a necessary and sufficient condition for 1-realizability of a function F on n variables as follows:

$$\begin{aligned}\sum_{i=0}^{n} a_i b_i &= \langle | \sum_{i=1}^{n} a_i x_{ji} + a_0 | \rangle \\ &= \langle |f(X_j)| \rangle; \quad f(X_j) \neq 0 \text{ for all } j.\end{aligned} \tag{6.8–1}$$

It can be shown that if F is not 1-realizable by the set of a_i, then

$$\begin{aligned}\sum_{i=0}^{n} a_i b_i &< \langle | \sum_{i=1}^{n} a_i x_{ji} + a_0 | \rangle \\ &= \langle |f(X_j)| \rangle; \quad f(X_j) \neq 0\end{aligned} \tag{6.8–2}$$

From Equations (6.8–1) and (6.8–2) it follows that if there exists a set of a_i that realizes a given function F, then a functional I can be defined such that

$$I_{min} = \left[\langle | \sum_{i=1}^{n} a_i x_{ji} + a_0 | \rangle - \sum_{i=0}^{n} a_i b_i \right]_{\min} = 0;$$

$$f(X_j) \neq 0 \tag{6.8–3}$$

where the minimization of the functional is performed by adjustments of $\{a_i\}$.

An exact explicit solution of Equation (6.8–3) is formidable. But an explicit solution for approximate values of $\{a_i\}$ can be obtained from a set of $(n+1)$ equations by making an approximate power series approximation for $|f(X_j)|$.

Dertouzos has shown that, for a first-order approximation,

$$a_i = b_i$$

and for a second-order approximation,

$$\left[\frac{n(2{\cdot}8n-3{\cdot}8)}{n(n+1)\sqrt{n}}\right] a_i = b_i - \frac{4{\cdot}4}{n\sqrt{n}2^{2n}} a_i^3$$

$$(i = 0, 1, \ldots, n) \tag{6.8–4}$$

There are in general $n+1$ such uncoupled cubic equations, and solution for the $\{a_i\}$ yields a set of weights and threshold value. No guarantee is given that this weight-threshold vector will always work. In fact, it is expected that as the number of input variables increases, more terms must be used in the truncated power-series approximation.

Although this method is too involved for practical application, it seems to be the only method that has a drastically different approach to the problem. The detailed analysis and procedure can be found in Dertouzos[4, 5] and will not be covered here.

We shall illustrate this method by the same example as above again.

EXAMPLE

Let $y_i \in \{0, 1\}$, $x_i \in \{-1, 1\}$, $F' \in \{0, 1\}$, $F \in \{-1, 1\}$.

$F' = y_1y_2 + y_1y_3 + y_2y_3y_4$

TABLE V

Table of Combinations

j	x_{j1}	x_{j2}	x_{j3}	x_{j4}	$F'(X_j)$	$F(X_j)$
0	−1	−1	−1	−1	0	−1
1	−1	−1	−1	1	0	−1
2	−1	−1	1	−1	0	−1
3	−1	−1	1	1	0	−1
4	−1	1	−1	−1	0	−1
5	−1	1	−1	1	0	−1
6	−1	1	1	−1	0	−1
7	−1	1	1	1	1	1
8	1	−1	−1	−1	0	−1
9	1	−1	−1	1	0	−1

TABLE V (cont.)

10	1	−1	1	−1	1	1
11	1	−1	1	1	1	1
12	1	1	−1	−1	1	1
13	1	1	−1	1	1	1
14	1	1	1	−1	1	1
15	1	1	1	1	1	1

From Table V we obtain:

$$b_0 = \langle F(X_j)\rangle = -2$$

$$b_1 = \langle x_{j1}F(X_j)\rangle = 10$$

$$b_2 = \langle x_{j2}F(X_j)\rangle = 6$$

$$b_3 = \langle x_{j3}F(X_j)\rangle = 6$$

$$b_4 = \langle x_{j4}F(X_j)\rangle = 2$$

For a first-order approximation, we assume that $a_i = b_i$. It is found that Theorem 3.7.1, or the equation

$$A.B = \langle |f_A(X_j)|\rangle$$

is satisfied. Therefore F is realized by A and no second-order approximation needs to be tried.

Going back to weights and threshold value for binary variables $y_i \in \{0, 1\}$ and $F' \in \{0, 1\}$, we have

$$a'_1 = a_1 = b_1 = 10$$
$$a'_2 = a_2 = b_2 = 6$$
$$a'_3 = a_3 = b_3 = 6$$
$$a'_4 = a_4 = b_4 = 2$$

$$T' = \frac{1}{2}\left(T + \sum_{i=1}^{n} a_i\right) = \tfrac{1}{2}(2+24) = 13$$

Or

$$R[F'] = (10, 6, 6, 2; 13)$$

6.9 A MAP METHOD

Torng[20] developed a map method for the testing and realization of threshold functions. The map facilitates the generation of inequalities and the detection of contradictions, and also eliminates any preprocessing (such as positivizing and ordering) of a given function.

In defining the threshold function, in addition to the threshold, Torng introduced a constant input 1 which is assigned a weight a_0. Besides, the upper and lower boundaries T_u and T_l are used instead of a single threshold value T. Thus

$$F = 1, \quad \text{if} \quad f(X) = a_0 + \sum_{i=1}^{n} a_i x_i \geq T_u$$

$$F = 0, \quad \text{if} \quad f(X) = a_0 + \sum_{i=1}^{n} a_i x_i \leq T_l$$

The value of $f(X)$ corresponding to a vertex X is called a *composite level*, and is denoted by f_m, where m is the decimal equivalent of the vertex.

A *map* is used to depict the relationships between the f's. Since when n is large the map becomes quite complex, the map for $n = 4$ is considered as standard and is called a *module*. For $n > 4$, more than one module is used. For instance, when $n = 6$, there are 4 modules.

Each node in the map corresponds to a composite level, and consequently a vertex. Node k corresponds to f_k. Each link is characterized by a α_{jk}, defined as

$$\alpha_{jk} = f_j - f_k$$

It can be shown that for a function of n variables, there are only n independent α's. When $n = 4$, the independent α's are chosen as α_{10}, α_{21}, α_{42}, and α_{84}. Any composite level f_k has one or more expressions as the sum of another composite level f_j and one of the n independent α's. Thus in this way the map can be constructed.

Given any two nodes, f_j and f_k, and a path between them with links labelled as α_{mn}, α_{no}, ..., the following equality is obtained.

$$f_j \pm \alpha_{mn} \pm \alpha_{no} \pm \ldots = f_k$$

The $\pm$ sign is resolved by the following convention: If, in reaching f_k from f_j, a link is traversed in the direction of the arrow, + is used. Otherwise, − is used.

A module is shown in Fig. 4.

For a given function F of n variables, the nodes representing the vertices of F are circled. Consider our previous Example again, *viz.* $F = x_1x_2 + x_1x_3 + x_2x_3x_4$. The vertices of F are circled in Fig. 4.

If f_j corresponds to a circled node and f_k corresponds to an uncircled node,

$$f_j \geq T_u \quad \text{and} \quad f_k \leq T_l$$

Or

$$f_j - f_k \geq T_u - T_l = g$$

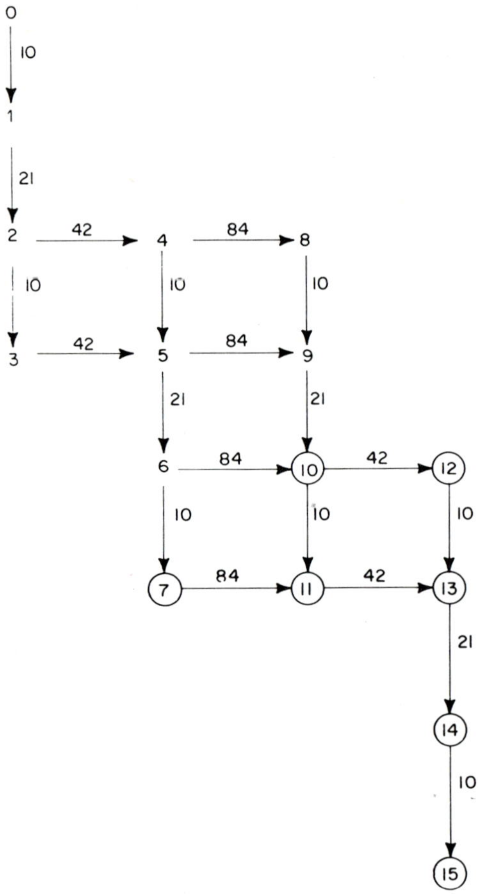

FIG. 4. A module or a map for $n = 4$.

Since $f_j - f_k = -(\pm\alpha_{mn} \pm \alpha_{no} \pm \ldots)$, we have

$$\alpha_{mn} \pm \alpha_{no} \pm \ldots \leq -g \qquad (6.9\text{–}1)$$

If f_j corresponds to an uncircled node and f_k corresponds to a circled node,

$$f_j \leq T_l \qquad \text{and} \qquad f_k \geq T_u$$

Or

$$f_j - f_k \leqslant -(T_u - T_l) = -g$$

Thus we have

$$\alpha_{mn} \pm \alpha_{no} \pm \ldots \geq g \qquad (6.9\text{–}2)$$

Inequalities of the types in (6.9–1) and (6.9–2) are generated for all pairs of adjacent nodes of which exactly one is circled. Each inequality thus generated contains one α and is called a simple inequality. This set of simple inequalities is the initial set of inequalities to be used.

Then the α's and the composite levels are tabulated, and values are assigned to the α's and the composite levels from left to right, by trial and error, guided by the set of existing inequalities. When a composite level corresponding to a vertex in F is equal to or less than one corresponding to a vertex in $\bar{F}$, a new inequality is generated and added to the existing set of inequalities, and the values of α's and the composite levels are to be readjusted.

If it is found that a newly introduced inequality is contradictory to the existing set of inequalities, then the given function F is not a threshold function. If no such contradiction exists, i.e. all the composite levels can be assigned values such that every composite level corresponding to a vertex in F is greater than any composite level corresponding to a vertex in $\bar{F}$, then F is a threshold function. The weights $\{a_i\}$ can be computed from the f's as follows:

$$
\begin{aligned}
a_0 &= f_0 \\
a_1 &= f_2^{n-1} - f_0 \\
a_2 &= f_2^{n-2} - f_0 \\
&\cdots \\
a_n &= f_1 - f_0
\end{aligned}
$$

The threshold value T can be assigned any value such that $T_l < T \leqslant T_u$, where T_u is the minimum value of composite levels corresponding to vertices in F, and T_l is the maximum value of composite levels corresponding to vertices in $\bar{F}$.

It is to be noted that, like the trial method of Gabelman[9] and the successive higher ordering method of Sheng and Hwa,[16, 17] this method also makes use of the differences between weights, as we can easily see that for $n = 4$, $\alpha_{84} = a_1 - a_2$, $\alpha_{42} = a_2 - a_3$, $\alpha_{21} = a_3 - a_4$, and $\alpha_{10} = a_4$. The features of this method, however, are that a map is used to help eliminate redundant inequalities, and that the α's and the weights can be negative. Thus the given function need not be canonical, nor positive in all the variables.

We shall illustrate this method by the same Example.

EXAMPLE

$F = x_1x_2 + x_1x_3 + x_2x_3x_4$

Since $n = 4$, only one module is needed, which is shown in Fig. 4. The nodes representing vertices in F are circled. From the map the following simple inequalities are obtained. Assuming $g = 1$, we have

$$\alpha_{84} \geq g = 1$$
$$\alpha_{21} \geq g = 1$$
$$\alpha_{10} \geq g = 1$$

This first assignment of the α's is as follows. Since α_{84}, α_{21}, and α_{10} appear in the simple inequalities, we assign

$$\alpha_{84} = \alpha_{21} = \alpha_{10} = 1$$

Since α_{42} does not appear in the set of simple inequalities, we assign

$$\alpha_{42} = 0$$

Since f_0 is taken as a reference, it may be assigned any value. Thus we assign

$$f_0 = 0$$

Then a chart is prepared as shown below. The first n columns list the α's and the remaining 2^n columns list the 2^n composite levels. A check mark ($\surd$) is placed on top of the columns representing vertices in F. The values for the composite levels are computed from the assigned values of f_0 and the α's with the help of the map.

TABLE VI

α_{10}	α_{21}	α_{42}	α_{84}	0	1	2	3	4	5	6	$\surd$ 7	8	9	$\surd$ 10	$\surd$ 11	$\surd$ 12	$\surd$ 13	$\surd$ 14	$\surd$ 15
1	1	0	1	0	1	2	3	2	3	4	5	3	4	5	6	5	6	7	8

Since in the first row of values shown in Table VI, every composite level corresponding to a vertex in F is greater than any composite level corresponding to a vertex in $\bar{F}$, F is a threshold function and is already realized by the assigned and computed values. No more readjustment is required.

Thus we have

$$a_0 = f_0 = 0$$
$$a_1 = f_8 - f_0 = 3 - 0 = 3$$
$$a_2 = f_4 - f_0 = 2 - 0 = 2$$
$$a_3 = f_2 - f_0 = 2 - 0 = 2$$
$$a_4 = f_1 - f_0 = 1 - 0 = 1$$
$$T = 5$$

Or

$$R[F] = (3, 2, 2, 1; 5)$$

REFERENCES

1. Akers, S. B. (1961). Threshold logic and two-person zero-sum games, in "Switching Circuit Theory and Logical Design", *A.I.E.E. Special Publication* **S134,** 27–33.
2. Coates, C. L., and Lewis, P. M. II. (1961). Linearly separable switching functions, *J. Franklin Inst.* **272,** 360–410.
3. Coates, C. L., Kirchner, R. B., and Lewis, P. M. II. (1962). A simplified procedure for the realization of linearly separable switching functions, *I.R.E. Trans. electronic Comput.* EC**11,** 447–458.
4. Dertouzos, M. L. (1964). An approach to single-threshold-element synthesis. *I.E.E.E. Trans. electronic Comput.* EC**13,** 519–528.
5. Dertouzos, M. L. (1965). "Threshold logic", M.I.T. Press, U.S.A.
6. Einhorn, S. N. (1961). The use of the simplex algorithm in the mechanization of Boolean switching functions by means of magnetic cores, *I.R.E. Trans. electronic Comput.* EC**10,** 615–622.
7. Elgot, C. C. (1961). Truth functions realizable by single threshold organs, in "Switching Circuit Theory and Logical Design", *A.I.E.E. Special Publication* **S134,** 225–245.
8. Fan, Ky (1967). On system of linear inequalities, Annals of Mathematics Studies, No. 38, pp. 99–156, Princeton University Press, U.S.A.
9. Gabelman, I. J. (1962). The synthesis of Boolean functions using a single threshold element, *I.E.E.E. Trans. electronic Comput.* EC**11,** 639–642.
10. Gaston, C. A. (1962). A simple test for linear separability, *I.E.E.E. Trans. electronic Comput.* EC**12,** 134–135.
11. Hu, S. T. (1965). "Threshold logic", University of California Press, U.S.A.
12. McNaughton, R. (1961). Unate truth functions, *I.R.E. Trans. electronic Comput.* EC**10,** 1–6.
13. Minnick, R. C. (1961). Linear-input logic, *I.R.E. Trans. electronic Comput.* EC**10,**, 6–16.
14. Muroga, S., Toda, I., and Takasu, S. (1961). Theory of majority decision elements, *J. Franklin Inst.* **271,** 376–418.
15. Roy, K. K., and Choudhury, A. K. (1966). A note on testing and realization of threshold functions, *I.E.E.E. Trans. electronic Comput.* EC**15,** 242–244.
16. Sheng, C. L. (1964). A method for testing and realization of threshold functions, *I.E.E.E. Trans. electronic Comput.* EC**13,** 232–239.
17. Sheng, C. L., and Hwa, H. R. (1966). The successive higher ordering method for the testing and realization of threshold functions, *I.E.E.E. Trans. electronic Comput.* EC**15,** 212–220.
18. Stram, O.B. (1961). Arbitrary Boolean functions of *N* variables realizable in terms of threshold devices, *Proc. Inst. Radio Engrs* **49,** 210–220.
19. Stram, O. B. (1961). The profile technique for the design of threshold device logic, in "Switching Circuit Theory and Logical Design", *A.I.E.E. Special Publication* **S134,** 47–54.
20. Torng, H. C. (1966). An approach for the realization of linearly separable switching functions, *I.E.E.E. Trans. electronic Comput.* EC**15,** 14–20.
21. Vajda, S. (1956). The theory of games and linear programming, John Wiley and Sons, Inc., New York, U.S.A.
22. Winder, R. O. (1961). Single stage threshold logic, in "Switching Circuit Theory and Logical Design", *A.I.E.E. Special Publication* **S134,** 321–332.
23. Winder, R. O. (1962). "Threshold logic", Doctoral dissertation, Mathematics Department, Princeton University, U.S.A.

CHAPTER 7

The Successive Higher Ordering Method for Testing and Realization

7.1 EXPLANATION OF THE METHOD

In this chapter we shall study the *successive higher ordering method* for testing and realization,[3,4] and its extension to functions with don't care vertices.[5] It will be seen that this method is particularly convenient for minimal-integral realization.

To begin with, we assume that the given function F of n variables is canonical and 2-monotonic, as there is no loss of generality. However, in applying this method, it is not necessary to know *a priori* that F is canonical and 2-monotonic, because the primary and successively higher orderings of the weights can be revealed by the method in the course of solution. As shown in its application to functions with don't care vertices, it is only necessary to know that the function is positive and 1-monotonic, so that all the weights may be assumed to be positive. However, since a threshold function is necessarily 2-monotonic, it is completely legitimate to test for 2-monotonicity and to find the ordering of weights before the application of the method. By so doing we can obtain the set of canonically minimal vertices in F, and the set of canonically maximal vertices in $\bar{F}$, thus greatly reducing the number of inequalities.

If the given function F of n variables is not symmetric in any subset of variables, then, by Theorem 5.3.5, we have

$$a_1 > a_2 > \dots > a_n > 0$$

If F is symmetric in one or more subsets of variables, we can have two different orderings of weights, depending on whether we require minimal-integral realization or not.

If minimal-integral realization is required, the weights of the variables in a symmetric subset are denoted by different symbols, and among these weights initially there exists no primary ordering. For instance, if F is symmetric in x_2 and x_3, we have

$$a_1 > \begin{matrix} a_2 \\ a_3 \end{matrix} > a_4 > \dots > a_n > 0$$

If minimal-integral realization is not required, then we can assume that all the m variables in a symmetric subset have the same weight, or the same symbol a_i can be used to denote the weight of each variable in a symmetric subset of m variables, because by Theorem 5.3.5, it is admissible to associate the variables in a symmetric subset with the same weight. Thus, if F is symmetric in x_2 and x_3, we can denote the weights of both x_2 and x_3 by a_2 and have

$$a_1 > a_2 (= a_3) > a_4 > \ldots > a_n > 0$$

In this case the same subscript can be used for the m variables in a symmetric subset. A superscript j will be used to denote the order of symmetric variables in a term or terms. This notation is adopted to avoid confusion and because the product of j symmetric variables has the same weight as j times the weight of each variable, and therefore such a product, insofar as its weight is concerned, is equivalent to a variable raised to the jth power. Thus x_i^j stands for the sum of all the products of j symmetric variables ranked ith in weight in the ordering.

For example, if x_{1a}, x_{1b}, and x_{1c} are three variables in which a certain function F is symmetric, then $x_{1a}+x_{1b}+x_{1c}$ will be denoted by x_1, $x_{1a}x_{1b}+x_{1a}x_{1c}+x_{1b}x_{1c}$ will be denoted by x_1^2, and $x_{1a}x_{1b}x_{1c}$ will be denoted by x_1^3. In this way a function symmetric in one or more subsets of variables can be expressed in a neater and simpler form with fewer terms than the original expression. The number of unknowns will be reduced, and the process of testing and realization will be shortened and simplified.

Since for a threshold function the inequalities are used to solve for the coefficients or weights, the inequalities may be expressed as functions of A. Let

$$f_A(X_j) = g_j(A)$$

$$\text{and} \quad f_A(X_k) = g_k(A)$$

Using the set of canonically minimal vertices in F and the set of canonically maximal vertices in $\bar{F}$, we have the following inequalities.

$$\begin{aligned} &I_{p'}: g_j(A) \geq T, \qquad j = 1, \ldots, p' \\ &I_{q'}: g_k(A) < T, \qquad k = p'+1, \ldots, p'+q' \end{aligned} \tag{7.1–1}$$

Usually the inequalities are solved for the weights $a_1, \ldots, a_n$ of the n variables. Since the ordering of the weights is already known, we can express the weights $a_1, a_2, \ldots, a_n$ in terms of weight a_n and the differences between successive weights, $\Delta a_1 = a_1 - a_2$, $\Delta a_2 = a_2 - a_3$, $\ldots$, $\Delta a_{n-1} = a_{n-1} - a_n$, which will be called *incremental weights*. Then the inequalities have n new unknowns $\Delta a_1, \Delta a_2, \ldots, \Delta a_{n-1}$, and a_n. A comparison of the set of inequalities will reveal the relative magnitudes of all or some of these incremental weights, and consequently we can find an *ordering of incremental weights*. We shall call the ordering of weights themselves *primary ordering* and the ordering of incremental weights *secondary ordering*. To express the weights in terms of incremental weights, we have

$$\begin{aligned} a_1 &= \Delta a_1 + \Delta a_2 + \ldots + \Delta a_{n-1} + a_n \\ a_2 &= \Delta a_2 + \ldots + \Delta a_{n-1} + a_n \\ \ldots \\ a_{n-1} &= \Delta a_{n-1} + a_n \\ a_n &= a_n \end{aligned} \tag{7.1–2}$$

Thus we have changed the set of $p'+q'$ inequalities in the n unknowns $a_1, a_2, \ldots, a_n$ into a set of $p'+q'$ inequalities in the n unknowns $\Delta a_1, \Delta a_2, \ldots, \Delta a_{n-1}, a_n$, as follows.

$$I_{p'}: g_j(\Delta a_1, \ldots, a_n) \geq T - C \qquad \text{for } j = 1, 2, \ldots, p'$$

$$I_{q'}: g_k(\Delta a_1, \ldots, a_n) < T - C \qquad \text{for } k = p'+1, \ldots, p'+q'$$

where C is the sum of all terms common to all inequalities.

Now we can expand the $p'+q'$ inequalities into $p' \times q'$ inequalities. To save space, the expanded $p' \times q'$ inequalities can be written with the p' rows of $I_{p'}$ on the left side and with the q' rows of $I_{q'}$ on the right side and with a $>$ sign between them to indicate that each row on the left side is greater than each row on the right side, as shown in inequalities (7.1–3). For clarity, $g_1, \ldots, g_{p'}$ are relabelled as $g_{L1}, \ldots, g_{Lp'}$, respectively, and $g_{p'+1}, \ldots, g_{p'+q'}$ are relabelled as $g_{R1}, \ldots, g_{Rq'}$, respectively.

$$\begin{matrix} g_{L1}(\Delta a_1, \ldots, a_n) & & g_{R1}(\Delta a_1, \ldots, a_n) \\ \ldots & > & \\ g_{Lp'}(\Delta a_1, \ldots, a_n) & & g_{Rq'}(\Delta a_1, \ldots, a_n) \end{matrix} \tag{7.1–3}$$

Let

S_{Lj} = sum of coefficients of all the terms in g_{Lj}, where $j = 1, \ldots, p'$

S_{Rk} = sum of coefficients of all the terms in g_{Rk}, where $k = 1, \ldots, q'$.

Next we want to find successively *higher orderings of weights*, or to express the incremental weights again in terms of *higher order incremental weights*.

Consider Inequalities (7.1–3). Some have the coefficients such that

$$S_{Lj} > S_{Rk}$$

These inequalities do not give any useful information as to higher ordering. So we simply ignore them.

Some of the inequalities have the coefficients such that

$$S_{Lj} \leq S_{Rk}$$

These inequalities are useful, and we shall consider them in detail.

By cancelling out common terms on both sides of each inequality, we can obtain simpler inequalities of the following two types.

(1) Type 1: $S_{Lj} = 1$, and $S_{Rk} \geq 1$. For instance,

$$\Delta a_1 > \Delta a_3 \tag{7.1–4}$$

$$\Delta a_1 > \Delta a_3 + \Delta a_4 \tag{7.1–5}$$

For Inequality (7.1–4), we set

$$\Delta a_1 = \Delta^2 a_1 + \Delta a_3 \tag{7.1–6}$$

For Inequality (7.1–5), we set

$$\Delta a_1 = \Delta^2 a_1 + \Delta a_3 + \Delta a_4 \tag{7.1–7}$$

By substituting Equation (7.1–6) and (7.1–7) into Inequalities (7.1–4) and (7.1–5), respectively, we obtain

$$\begin{gathered} \Delta^2 a_1 + \Delta a_3 > \Delta a_3, \\ \text{or} \\ \Delta^2 a_1 > 0 \end{gathered} \tag{7.1–8}$$

$$\begin{gathered} \Delta^2 a_1 + \Delta a_3 + \Delta a_4 > \Delta a_3 + \Delta a_4 \\ \text{or} \\ \Delta^2 a_1 > 0 \end{gathered} \tag{7.1–9}$$

Inequalities (7.1–8) and (7.1–9) are redundant, because we assume every weight and every incremental weight of any order to be positive. Now by substituting Equations (7.1–6) and (7.1–7) back into Inequalities (7.1–3), owing to the redundant nature of Inequalities (7.1–8) and (7.1–9), and owing to the fact that we add at least one to the sum of coefficients of certain rows having Δa_1, which are more likely to be on the left side than on the right side, we decrease the number of inequalities with $S_{Li} \leq S_{Rj}$.

In this way we continually search for higher-order incremental weights and substitute them back into the inequalities. The higher-order incremental weights are not assigned any fixed values until the end of the search process so that they are open to still higher ordering.

We shall call relations given by Inequalities (7.1–4) and (7.1–5) *explicit relations.*

(2) Type 2: $S_{Li} = k \geq 2$, and $S_{Rj} \geq k$. For instance,

$$\Delta a_1 + \Delta a_2 > \Delta a_4 + \Delta a_5 \tag{7.1–10}$$

$$\Delta a_1 + \Delta a_2 > \Delta a_4 + \Delta a_5 + \Delta a_6 \tag{7.1–11}$$

We shall call relations given by Inequalities (7.1–10) and (7.1–11) *implicit relations.*

In an implicit relation, there is at least one incremental weight on the left side that is greater than a certain incremental weight on the right side, although this relation may not appear as an explicit relation. For instance, if Inequality (7.1–10) is correct, then at least one of the following four relations must be correct.

$$\Delta a_1 > \Delta a_4, \quad \Delta a_1 > \Delta a_5, \quad \Delta a_2 > \Delta a_4, \quad \Delta a_2 > \Delta a_5 \qquad (7.1\text{–}12)$$

We shall call an explicit relation extracted from an implicit relation such as shown by Inequality (7.1–12) an *extracted explicit relation.*

In using an implicit relation, it is important to choose the correct extracted explicit relation. However, there is no systematic method to determine which is the correct extracted explicit relation. Fortunately, in the beginning of the testing process, there are always some explicit relations available. As long as there are explicit relations available, we do not use implicit relations. We resort to implicit relations only when there is no longer any explicit relation left. As shown by the example in Section 7.4, there is no difficulty in choosing the correct extracted explicit relation. As a general rule, we can choose an extracted explicit relation $\Delta a_i > \Delta a_j$ in such a way that Δa_i appears most frequently on the left side of the inequalities and Δa_j appears most frequently on the right side of the inequalities.

Once the extracted explicit relation is chosen, the setting of higher order incremental weight and the substitution of this higher order incremental weight back into the inequalities are the same as those for explicit relations.

If we continue this process of finding higher ordering incremental weights and substituting back into Inequalities (7.1–3), eventually we shall arrive at either one of the following two results:

(1) $S_{Lj} > S_{Rk}$ for all $j = 1, \ldots, p'$ and $k = 1, \ldots, q'$

Then the given function is a threshold function, because if we assign any same integer to every incremental weight of any order in the final set of inequalities, all the inequalities will be satisfied. Thus the minimal integral assignment is simply to assign unity to every incremental weight of any order in the final set of inequalities.

(2) There is some contradiction or contradictions manifested as $0 >$ a sum of one or more incremental weights of some orders.

Then the given function is not a threshold function.

7.2 THEOREMS SUPPORTING THE METHOD

THEOREM 7.2.1. A necessary and sufficient condition for a given function F to be a threshold function is that, if only explicit relations or correct extracted explicit relations from implicit relations are used, the successive substitution of higher order incremental weights will result in a set of inequalities such that

$$S_{Lj} > S_{Rk}, \qquad \text{for all } j \text{ and } k.$$

PROOF: *Necessity.* If the given function F is a threshold function, there must exist a minimal-integral assignment of weights. So we shall consider integral weights only. For the minimal-integral assignment, there must exist integral incremental weights up to certain orders.

Since Inequalities (7.1–3) must be satisfied, the higher order incremental weight obtained from an explicit relation or from a correct extracted explicit relation taken from an implicit relation must be a positive integer being at least unity.

Since F is a threshold function, there exists no contradiction. Therefore at any stage of the testing process, there are only two possibilities:

(1) $S_{Lj} > S_{Rk}$ for all j and k.
(2) $S_{Lj} \leq S_{Rk}$ for some j and k.

For case 1, there are no explicit and implicit relations, but Theorem 7.2.1 is already satisfied.

For case 2, since $S_{Lj} \leq S_{Rk}$ for some j and k, there must be at least one explicit or implicit relation, and therefore the process of successive higher ordering can be carried further on.

If the situation $S_{Lj} \leq S_{Rk}$ for some j and k continues to the end when there is only one inequality left, then by setting the left side equal to the right side plus a higher ordering incremental weight, we obtain $S_{Lj} > S_{Rk}$ for all j and k. Thus Theorem 7.2.1 is also satisfied.

Now we shall prove that the process of successive higher ordering is convergent.

For $\Delta a_i > \Delta a_j$, we set $\Delta a_i = \Delta^2 a_i + \Delta a_j$. Since all the incremental weights are positive integers,

$$\Delta a_i > \Delta^2 a_i$$

or in general

$$\Delta^k a_i > \Delta^{k+1} a_i$$

Thus each higher order incremental weight is smaller than the corresponding lower order incremental weight, or the successive higher order incremental weights will decrease monotonically. Since there exists an upper bound for each weight and for the sum of the weights, for a minimal-integral assignment each weight must be finite. The successive higher order incremental weights will all decrease to unity in a finite number of steps. Therefore the process is convergent.

Sufficiency. If $S_{Lj} > S_{Rk}$ for all j and k, by setting every incremental weight in the final set of inequalities to unity, all the inequalities are reduced to $S_{Lj} > S_{Rk}$, and thus are all satisfied. Therefore F must be a threshold function. Q.E.D.

COROLLARY 7.2.1–1. A necessary and sufficient condition for a given function F to be not a threshold function is that any successive substitution of higher order incremental weights will eventually result in a contradiction of the form

$$0 > \text{A sum of one or more incremental weights of some orders} \quad (7.2\text{–}1)$$

PROOF: *Necessity.* It can be shown that any contradiction can be reduced to the form of Inequality (7.2–1). If the given function F is not a threshold function, and if at any stage of the process no contradiction appears, then either (1) $S_{Lj} > S_{Rk}$ for all j and k, or (2) $S_{Lj} \leq S_{Rk}$ for some j and k. For case (1), according to Theorem 7.1.1, F is a threshold function. This contradicts the assumption. For case (2), the process can be carried further on. Again according to Theorem 7.1.1, when there is only one inequality left, we can obtain $S_{Lj} > S_{Rk}$ for all j and k, and F is also a threshold function, again contradicting the assumption.

The sufficiency of this corollary is obvious. Q.E.D.

In spite of Theorem 7.2.1, theoretically there is still some difficulty in applying this method. The difficulty lies essentially in the implicit relations. If at each stage there is at least one explicit relation available, then we can use explicit relations exclusively and simply ignore the implicit relations. Since each explicit relation represents a condition that must be satisfied, whenever we encounter a contradiction, it is implied that the given function is not a threshold function. When there is no explicit relation available, we must extract an explicit relation from an implicit relation. If a correct extraction is made, the situation is similar to that when only explicit relations are used. But a wrong extraction may lead to a contradiction even if the given function is a threshold function. For instance, suppose

$$\Delta a_1 + \Delta a_3 > \Delta a_2 + \Delta a_4$$

and

$$\Delta a_1 = 1, \Delta a_2 = 2, \Delta a_3 = 3, \Delta a_4 = 1$$

If we choose $\Delta a_1 > \Delta a_2$ and set $\Delta a_1 = \Delta^2 a_1 + \Delta a_2$, then $\Delta^2 a_1$ would be negative, and a contradiction will result. Therefore once we extract explicit relations from implicit relations, a contradiction does not imply that the given function F is not a threshold function. F is not a threshold function iff every extracted explicit relation from an implicit relation eventually results in a contradiction. So to use an implicit relation, theoretically we have to try all the possible extracted explicit relations therein. In this sense, trial and error is still not completely avoidable.

However, practically it is seldom necessary to resort to implicit relations, except in very unusual cases.

THEOREM 7.2.2. If a given function F is a threshold function, the assignment made by setting each incremental weight of any order to unity in the set of inequalities where $S_{Lj} > S_{Rk}$ for all j and k is a minimal-integral assignment.

PROOF: For integral assignment, each incremental weight should be at least unity. Before the set of inequalities with $S_{Lj} > S_{Rk}$ for all j and k is reached, such an assignment does not satisfy all the inequalities, because for some inequalities, $S_{Lj} \leq S_{Rk}$. When such a set of inequalities is reached, there are no more explicit and/or implicit relations left. Therefore once this set of inequalities is reached, no more higher ordering is possible, or this set of inequalities is unique in so far as the requirement of higher ordering of weights is concerned. Since unity is the smallest integer that can be assigned to any incremental weight of any order, this assignment must be minimal-integral. Q.E.D.

7.3 PROCEDURE OF TESTING AND REALIZATION

The procedures of testing for 1-realizability and of realization are combined together, because they are closely related.

Step 1: Checking the given function F for unateness and 2-monotonicity.

Step 2: Finding the complementary function $\bar{F}$.

Step 3: Reducing F and $\bar{F}$ to minimum sum-of-product form (or irredundant normal disjoint form).

Step 4: Primary ordering.

$$a_1 > a_2 > \ldots > a_n > 0$$

Step 5: Inequalities in A.

Write p' inequalities of the form $g_j(A) \geq T, j = 1, \ldots, p'$, corresponding to the set of canonically minimal vertices in F, and q' inequalities of the form $g_k(A) < T, k = p'+1, \ldots, p'+q'$, corresponding to the set of canonically maximal vertices in $\bar{F}$. Thus we have

$$\begin{aligned} I_{p'}: g_j(A) &\geq T \qquad \text{for } j = 1, \ldots, p' \\ I_{q'}: g_k(A) &< T \qquad \text{for } k = p'+1, \ldots, p'+q' \end{aligned} \tag{7.1–1}$$

Step 6: Substitution of incremental weights and reduction.

$$
\begin{aligned}
a_1 &= \Delta a_1 + \Delta a_2 + \ldots + \Delta a_{n-1} + a_n \\
a_2 &= \Delta a_2 + \ldots + \Delta a_{n-1} + a_n \\
&\ldots \\
a_{n-1} &= \Delta a_{n-1} + a_n \\
a_n &= a_n
\end{aligned}
\qquad (7.1\text{–}2)
$$

Substitute Equation (7.1–2) into Inequalities (7.1–1). Find C, the sum of all the terms common to all inequalities. Subtract C from each inequality. Then the inequalities become

$$
\left.
\begin{aligned}
&I_{p'}: g_1(\Delta a_1, \Delta a_2, \ldots, a_n) \\
&\ldots \\
&g_{p'}(\Delta a_1, \Delta a_2, \ldots, a_n)
\end{aligned}
\right\} \geq T - C
$$

$$
\left.
\begin{aligned}
&I_{q'}: g_{p'+1}(\Delta a_1, \Delta a_2, \ldots, a_n) \\
&\ldots \\
&\quad g_{p'+q'}(\Delta a_1, \Delta a_2, \ldots, a_n)
\end{aligned}
\right\} < T - C
$$

Step 7: "Expansion" of inequalities in tabular form.

$$
\begin{array}{lcl}
g_{L1}(\Delta a_1, \Delta a_2, \ldots, a_n) & & g_{R1}(\Delta a_1, \Delta a_2, \ldots, a_n) \\
\ldots & > & \ldots \\
g_{Lp'}(\Delta a_1, \Delta a_2, \ldots, a_n) & & g_{Rq'}(\Delta a_1, \Delta a_2, \ldots, a_n)
\end{array}
\qquad (7.1\text{–}3)
$$

Find S_{Lj}, $j = 1, \ldots, p'$, and S_{Rk}, $k = 1, \ldots, q'$, for the inequalities and list them on the left and right sides of the table, respectively.

Step 8: Comparison of S_{Lj} and S_{Rk} for all the rows in Inequalities (7.1–3).

(a) If $S_{Lj} > S_{Rk}$ for all j and k, the given function F is a threshold function. Proceed to Step 11.

(b) If $S_{Lj} \leq S_{Rk}$ for some j and k, proceed to Step 9.

Step 9: Check for contradiction.

Check each inequality with $S_{Lj} \leq S_{Rk}$ to see if it can be reduced to the form $0 >$ A sum of one or more incremental weights of some orders.

(a) If one or more inequalities can be reduced to the above form, the given function F is not a threshold function. Stop.

(b) If no inequality can be reduced to the above form, proceed to Step 10.

Step 10: Higher ordering of weights.

Of the $p' \times q'$ inequalities, ignore those with $S_{Lj} > S_{Rk}$. Consider only those inequalities with $S_{Lj} \leq S_{Rk}$. Compare every left-side row with every right-side row.

(a) If there are one or more explicit relations, express the left-side terms as follows:

$$\text{If} \quad \Delta a_i > \Delta a_j, \text{ set } \Delta a_i = \Delta^2 a_i + \Delta a_j \tag{7.3–1}$$

$$\text{If} \quad \Delta a_i > \Delta a_j + \Delta a_k + \ldots, \text{ set } \Delta a_i = \Delta^2 a_i + \Delta a_j + \Delta a_k + \ldots \tag{7.3–2}$$

Substitute (7.3–1), (7.3–2), etc. back into inequalities (7.1–3). Subtract all new common terms, if any. Delete all new redundant inequalities, if any. Go back to Step 8.

(b) If there are no explicit relations, choose extracted explicit relation or relations of the form $\Delta a_i > \Delta a_j$ from implicit relations. Extracted explicit relations are to be chosen in such a way that Δa_i appears most frequently on the left side and Δa_j appears most frequently on the right side of the inequalities. Set

$$\Delta a_i = \Delta^2 a_i + \Delta a_j \tag{7.3–1}$$

Substitute (7.3–1) back into inequalities (7.1–3). Subtract all new common terms, if any. Delete all new redundant inequalities, if any. Go back to Step 8.

Step 11: Assigning of weights and threshold value-realization. Assign unity to all the incremental weights of any order in the final set of inequalities (7.1–3). Using (7.1–2), (7.3–1) and (7.3–2), obtain incremental weights of all orders and the original weights. Substitute all weights into inequalities (7.1–1). Set

$$T = \text{minimum } g_j(A), \text{ for } j = 1, \ldots, p'.$$

Then the realization of the given function F is

$$R[F] = (a_1, a_2, \ldots, a_n; T).$$

7.4 EXAMPLE

$$\begin{aligned} F(x_1, x_2, x_3, x_4, x_5, x_6, x_7, x_8) \\ = x_1x_2 + x_1x_3 + x_1x_4 + x_1x_5 + x_1x_6x_7 + x_1x_6x_8 + x_2x_3x_4 + x_2x_3x_5 + \\ x_2x_3x_6 + x_2x_3x_7 + x_2x_3x_8 + x_2x_4x_5 + x_2x_4x_6 + x_2x_4x_7 + x_2x_4x_8 + \\ x_2x_5x_6 + x_2x_5x_7 + x_2x_5x_8 + x_2x_6x_7x_8 + x_3x_4x_5 + x_3x_4x_6 + \\ x_3x_4x_7x_8 + x_3x_5x_6 + x_3x_5x_7x_8 + x_4x_5x_6x_7 + x_4x_5x_6x_8 + \\ x_4x_5x_7x_8. \end{aligned}$$

F is symmetric in variables x_4 and x_5, and also symmetric in variables x_7 and x_8. We shall work out this example in detail, with the eight weights denoted by eight different symbols.

Step 1: F is unate and also 2-monotonic.

Step 2:

$$
\begin{aligned}
F = {} & \bar{x}_1\bar{x}_2\bar{x}_3\bar{x}_4+\bar{x}_1\bar{x}_2\bar{x}_3\bar{x}_5+\bar{x}_1\bar{x}_2\bar{x}_3\bar{x}_6\bar{x}_7+\bar{x}_1\bar{x}_2\bar{x}_3\bar{x}_6\bar{x}_8+\bar{x}_1\bar{x}_2\bar{x}_3\bar{x}_7\bar{x}_8\\
&+\bar{x}_1\bar{x}_2\bar{x}_4\bar{x}_5+\bar{x}_1\bar{x}_2\bar{x}_4\bar{x}_6\bar{x}_7+\bar{x}_1\bar{x}_2\bar{x}_4\bar{x}_6\bar{x}_8+\bar{x}_1\bar{x}_2\bar{x}_5\bar{x}_6\bar{x}_7\\
&+\bar{x}_1\bar{x}_2\bar{x}_5\bar{x}_6\bar{x}_8+\bar{x}_1\bar{x}_3\bar{x}_4\bar{x}_5\bar{x}_6+\bar{x}_1\bar{x}_3\bar{x}_4\bar{x}_5\bar{x}_7+\bar{x}_1\bar{x}_3\bar{x}_4\bar{x}_5\bar{x}_8\\
&+\bar{x}_1\bar{x}_3\bar{x}_4\bar{x}_6\bar{x}_7\bar{x}_8+\bar{x}_1\bar{x}_3\bar{x}_5\bar{x}_6\bar{x}_7\bar{x}_8+\bar{x}_1\bar{x}_4\bar{x}_5\bar{x}_6\bar{x}_7\bar{x}_8+\bar{x}_2\bar{x}_3\bar{x}_4\bar{x}_5\bar{x}_6\\
&+\bar{x}_2\bar{x}_3\bar{x}_4\bar{x}_5\bar{x}_7\bar{x}_8.
\end{aligned}
$$

Step 3: F and $\bar{F}$ are already in sum-of-product form.

Step 4:

$$a_1 > a_2 > a_3 > \begin{matrix} a_4 \\ a_5 \end{matrix} > a_6 > \begin{matrix} a_7 \\ a_8 \end{matrix} > 0$$

Step 5: The irredundant inequalities $I_{p'}$ and $I_{q'}$ are given in Table VII.

TABLE VII

	j	a_1	a_2	a_3	a_4	a_5	a_6	a_7	a_8			k	a_1	a_2	a_3	a_4	a_5	a_6	a_7	a_8	
	1	1			1							1	1					1			
	2	1				1						2	1						1	1	
	3	1					1	1				3		1	1						
	4	1					1		1			4		1				1	1		
	5		1		1			1				5		1				1		1	
g_{Lj}	6		1		1				1	$\geq T$	g_{Rk}	6			1	1			1		$< T$
	7		1			1		1				7			1	1				1	
	8		1			1			1			8			1		1		1		
	9		1				1	1	1			9			1		1			1	
	10			1	1		1					10			1			1	1	1	
	11			1		1	1					11				1	1	1			
	12				1	1		1	1												

Steps 6 and 7: Set

$$
\begin{aligned}
a_1 &= \Delta a_1+\Delta a_2+\Delta a_3+\Delta a_4+\Delta a_6+a_7\\
a_2 &= \Delta a_2+\Delta a_3+\Delta a_4+\Delta a_6+a_7\\
a_3 &= \Delta a_3+\Delta a_4+\Delta a_6+a_7\\
a_4 &= \Delta a_4+\Delta a_6+a_7\\
a_5 &= \Delta a_5+\Delta a_6+a_7\\
a_6 &= \Delta a_6+a_7\\
a_7 &= a_7\\
a_8 &= a_8.
\end{aligned}
\qquad (7.4\text{–}1)
$$

TABLE VIII

j	Δa_1	Δa_2	Δa_3	Δa_4	Δa_5	Δa_6	a_7	a_8	S_{Lj}
1	1	1	1	1		1			5
2	1	1	1		1	1			5
3	1	1	1			1			5
4	1	1	1			1		1	5
5		1	1	1		1			5
6		1	1	1		1		1	5
g_{Lj} 7		1	1		1	1	1		5
8		1	1		1	1		1	5
9		1	1			1	1	1	5
10			1	1		2	1		5
11			1		1	2	1		5
12					1	1	1	1	4

$> g_{Rk}$

k	Δa_1	Δa_2	Δa_3	Δa_4	Δa_5	Δa_6	a_7	a_8	S_{Rk}
1	1	1	1			1			4
2	1	1	1					1	4
3		1	2	1		1			5
4		1	1			1	1		4
5		1	1			1		1	4
6			1	1		1	1		4
7		1	1			1		1	4
8			1		1	1	1		4
9			1		1	1		1	4
10			1			1	1	1	4
11					1	2	1		4

TABLE IX

j	$\Delta^2 a_1$	Δa_2	Δa_3	Δa_4	$\Delta^2 a_5$	Δa_6	Δa_7	a_8	S_{Lj}
1	1	1		1		1			4
2	1	1	1		1	1			5
3	1	1	1			1	1		5
4	1	1	1			1		1	5
5		1		1		1	1		4
g_{Lj} 6		1		1		1		1	4
7		1	1		1	1	1		5
8		1	1		1	1		1	5
9		1	1			1	1	1	5
10				1		2	1		4
11			1		1	2	1		5
12			1		1	1	1	1	5

$> g_{Rk}$

k	$\Delta^2 a_1$	Δa_2	Δa_3	Δa_4	$\Delta^2 a_5$	Δa_6	Δa_7	a_8	S_{Rk}
1	1	1				1			3
2	1	1	1					1	4
3		1		1		1			3
4		1				1	1		3
5		1				1		1	3
6				1		1	1		3
7				1		1		1	3
8			1		1	1	1		4
9			1		1	1		1	4
10			1			1	1	1	4
11					1	2	1		4

TABLE X

j	$\Delta^2 a_1$	Δa_2	Δa_3	$\Delta^2 a_4$	$\Delta^2 a_5$	Δa_6	Δa_7	Δa_8	S_{Lj}
1	1	1		1	1	1			5
2	1	1	1		1	1			5
3	1	1	1			1	1		5
4	1	1	1			1		1	5
5		1		1	1	1	1		5
6		1		1	1	1		1	5
g_{Lj} 7		1	1		1	1	1		5
8		1	1		1	1		1	5
9		1	1			1	1	1	5
10				1	1	2	1		5
11			1		1	2	1		5
12			1		1	1	1	1	5

$> g_{Rk}$

k	$\Delta^2 a_1$	Δa_2	Δa_3	$\Delta^2 a_4$	$\Delta^2 a_5$	Δa_6	Δa_7	Δa_8	S_{Rk}
1	1	1				1			3
2	1	1	1					1	4
3		1		1	1	1			4
4		1				1	1		3
5		1				1		1	3
6				1	1	1	1		4
7				1	1	1		1	4
8			1		1	1	1		4
9			1		1	1		1	4
10			1			1	1	1	4
11					1	2	1		4

Substituting (7.4–1) into Table VII and subtracting the common terms $C = \Delta a_4 + \Delta a_8 + 2a_7$ from each inequality, we obtain Table VIII.

Step 8: Condition *b* is satisfied.

Step 9: Condition *b* is satisfied.

Step 10: From Table VIII, we find

$$\begin{array}{ll} g_{L1} - g_{R3} & \Delta a_1 > \Delta a_3 \\ g_{L5} - g_{R3} & a_7 > \Delta a_3 \\ g_{L6} - g_{R3} & a_8 > \Delta a_3 \\ g_{L12} - g_{R8} & a_8 > \Delta a_3 \\ g_{L12} - g_{R9} & a_7 > \Delta a_3 \\ g_{L12} - g_{R10} & \Delta a_5 > \Delta a_3 \\ g_{L12} - g_{R11} & a_8 > \Delta a_6. \end{array}$$

Condition *a* is satisfied. Set

$$\begin{aligned} \Delta a_1 &= \Delta^2 a_1 + \Delta a_3 \\ \Delta a_5 &= \Delta^2 a_5 + \Delta a_3 \\ a_7 &= \Delta a_7 + \Delta a_3 \\ a_8 &= \Delta a_8 + \Delta a_3. \end{aligned} \tag{7.4–2}$$

Substituting (7.4–2) back into Table VIII and subtracting the new common term $C = 2\Delta a_3$, we obtain Table IX.

Step 8: Condition *b* is satisfied.

Step 9: Condition *b* is satisfied.

Step 10: From Table IX, we find

$$g_{L10} - g_{R11} \quad \Delta a_4 > \Delta^2 a_5.$$

Condition *a* is satisfied. Set

$$\Delta a_4 = \Delta^2 a_4 + \Delta^2 a_5. \tag{7.4–3}$$

Substituting (7.4–3) back into Table IX, we obtain Table X.

Step 8: Condition *a* is satisfied. So we proceed to Step 11.

Step 11: Assign

$$\Delta^2 a_1 = \Delta a_2 = \Delta a_3 = \Delta^2 a_4 = \Delta^2 a_5 = \Delta a_6 = \Delta a_7 = \Delta a_8 = 1.$$

Substituting these values back into (7.4–1), (7.4–2), (7.4–3), and Table VII we obtain

$$\begin{aligned} &a_1 = 9, a_2 = 7, a_3 = 6, a_4 = 5, a_5 = 5, \\ &a_6 = 3, a_7 = 2, a_8 = 2, \text{ and } T = 14. \end{aligned}$$

Therefore,

$$R[F] = (9, 7, 6, 5, 5, 3, 2, 2; 14).$$

If in the very beginning symmetric variables are assumed to have equal weights, or $a_4 = a_5$ and $a_7 = a_8$, then the procedure will be greatly shortened. It is found that the solution can be obtained with only one substitution of incremental weights after secondary ordering, or with two substitutions of incremental weights altogether, as compared with three substitutions in the above solution. For this particular function, the solution obtained with symmetric variables having equal weights is the same as the above solution, or is a minimal-integral realization, although this is not necessarily true in general.

It is of interest to note that, in the solution of this example, no implicit relations are used. Therefore, there is no trial and error at all. It should be emphasized that as long as only explicit relations and correct extracted explicit relations are used, if more relations than required are available, any relations chosen will work and will result in a minimal-integral realization. For instance, in the above example, if we set $a_8 = \Delta a_8 + \Delta a_6$ instead of $\Delta a_8 + \Delta a_3$, the same realization will be obtained eventually.

7.5 TESTING AND REALIZATION OF THRESHOLD FUNCTIONS WITH DON'T CARE VERTICES

The problem of testing and realization of threshold functions *with don't care vertices* has been studied by Winder,[6] Coates and Lewis II,[1] Hu,[2] Sheng and Hwa,[5] etc. It is found that the successive higher ordering method can be very easily and conveniently applied to this problem. Stated more precisely, this problem is to test whether it is possible, and if possible, how, to realize a threshold function F such that F will contain a given function F_1 which may or may not be a threshold function, and such that F will be contained in $F_1 + F_\varphi$, where F_φ is the function representing all the don't care vertices.

In geometrical interpretation, the ordinary 1-realization problem is that, given a function F, it is required to find a hyperplane $\sum_{i=1}^{n} a_i x_i = T$ to separate F from $\bar{F}$, such that all the vertices in F will lie on one side of the hyperplane and all the vertices in $\bar{F}$ will lie on the other side. Now the problem of 1-realization with don't care vertices is that, given a function F_1, and a function F_φ representing all the don't care vertices, it is required to find a hyperplane to separate F_1 from $F_0 = \overline{(F_1 + F_\varphi)}$, such that all the vertices in F_1 will lie on one side of the hyperplane, and all the vertices in F_0 will lie on the other side, whereas it is immaterial whether the vertices in F_φ lie on which side of the hyperplane. In other words, for ordinary 1-realization, the hyperplane has to pass through the "gap", if any, between F and $\bar{F}$, while for 1-realization with don't care vertices, the hyperplane can pass through the space occupied by the don't care vertices. It is seen that there is more freedom in this latter case. If a given function is 1-realizable, there may exist one or more threshold functions which will each satisfy the requirement.

For the case with don't care vertices, we shall define 1-realizability as follows:

For a given function F_1 and another function F_ϕ representing all the don't care vertices, F_1 is said to be 1-realizable by a threshold function F if there exists a threshold function F such that

$$F_1 + F_\phi \supset F \supset F_1$$

F_1 and F_0 are said to be linearly separable by F iff F_1 is 1-realizable by a threshold function F.

It will be proved later that a necessary condition for 1-realizability is the existence of at least one pair of unate functions $\{F_{1U}\}$ such that $F_{1U} \supset F_1$ and $F_{0U} \supset F_0$. So before the application of the successive higher ordering method the given functions F_1 and $F_0 = \overline{(F_1 + F_\phi)}$ are first changed into unate functions. Once the pair or pairs of unate functions are formed, the successive higher ordering method can be applied directly to each pair, because the inequalities contain all the information on weights. Sometimes there may not exist a complete primary ordering. This simply means that the non-existing part of ordering is not a necessary condition to be satisfied by the weights. As long as the existing partial ordering is satisfied, the given function can be realized.

7.6 FORMATION OF UNATE FUNCTIONS

Before the application of the successive higher ordering method, we must positivize F_1 with respect to the variables $\{x_i^*\}$, i.e. to make F_1 positive in the variables $\{x_i^*\}$, because otherwise the signs of the weights are not known, and it is impossible to have comparison and ordering of the weights. If a function is positive in all the variables $\{x_i^*\}$, it is a unate function. So the first problem is to change F_1 and F_0 into unate functions.

THEOREM 7.6.1. A necessary condition for a given function F_1 to be 1-realizable is that there exists a unate function F_{1U} and a unate function F_{0U} such that

$$F_{1U} \supset F_1, \quad F_{0U} \supset F_0, \quad \text{and} \quad F_{1U}F_{0U} = 0 \tag{7.6–1}$$

PROOF: If F_1 is 1-realizable, then according to the definition, there exists a threshold function F such that

$$F_1 + F_\phi \supset F \supset F_1$$

$$\overline{F}_1 \supset \overline{F} \supset \overline{(F_1 + F_\phi)} = F_0 \tag{7.6–2}$$

and

$$F\overline{F} = 0$$

A threshold function is necessarily a unate function. Therefore F and $\bar{F}$ are both unate functions, and Equation (7.6–2) implies Equation (7.6–1).

Q.E.D.

THEOREM 7.6.2. For a function F_1, there exist a function $F_{1(x_i)} \supset F_1$ and positive in x_i such that if any other function $F'_{1(x_i)} \supset F_1$ and is positive in x_i, then

$$F'_{1(x_i)} \supset F_{1(x_i)}$$

PROOF: Set $F_1 = x_iP + \bar{x}_iQ + R$ in normal form, where P, Q, and R are each either 0 or a function independent of x_i. It is obvious that to make F_1 positive in x_i, we can add x_iQ to F_1. Let

$$F_{1(x_i)} = F_1 + x_iQ = x_iP + \bar{x}_iQ + R + x_iQ = x_iP + Q + R \quad (7.6\text{–}3)$$

Consider another function $F'_{1(x_i)} \supset F_1$ and positive in x_i.

$$F'_{1(x_i)} \supset x_iP + \bar{x}_iQ + R \quad (7.6\text{–}4)$$

Suppose $F'_{1(x_i)} \not\supset x_iQ$. Then there must exist a minterm x_iS (where S is the product of all variables except x_i) which is contained in x_iQ but not contained in $F'_{1(x_i)}$. But

$$\bar{x}_iS \subset \bar{x}_iQ \subset F'_{1(x_i)}$$

Therefore $F'_{1(x_i)}$ is not positive in x_i, which contradicts the assumption that $F'_{1(x_i)}$ is positive in x_i. Thus

$$F'_{1(x_i)} \supset x_iQ \quad (7.6\text{–}5)$$

From Equations (7.6–4) and (7.6–5), we have

$$F'_{1(x_i)} \supset (x_iP + \bar{x}_iQ + R) + x_iQ$$

or

$$F'_{1(x_i)} \supset F_{1(x_i)} \quad (7.6\text{–}6)$$

Q.E.D.

COROLLARY 7.6.2–1. For F_1, $F_{1(x_i)}$ is unique.

PROOF: Suppose $F'_{1(x_i)}$ is also the "smallest" function which is positive in x_i and contains F_1. Then

$$F_{1(x_i)} \supset F'_{1(x_i)} \quad (7.6\text{–}7)$$

From Equations (7.6–6) and (7.6–7), we obtain

$$F'_{1(x_i)} = F_{1(x_i)}$$

Q.E.D.

$F_{1(x_i)}$ is called the *minimum x_i-function* of F_1.

COROLLARY 7.6.2–2. The minimum x_i-function $F_{1(x_i)}$ of any function F_1 can be formed from F_1 by deleting all $\bar{x}_i$'s in the irredundant normal disjunctive form of F_1.

THEOREM 7.6.3. Given any function F_1 of n variables, for any subset of k variables $x_{i_1}, x_{i_2}, \ldots, x_{i_k}$, $k = 2, 3, \ldots, n$, there exists a function $F_1(x_{i_1}, \ldots, x_{i_k}) \supset F_1$ and positive in $x_{i_1}, x_{i_2}, \ldots, x_{i_k}$, formed by making F_1 positive in $x_{i_1}, x_{i_2}, \ldots, x_{i_k}$ successively, such that if $F'_{1(x_{i_1}, \ldots, x_{i_k})} \supset F_1$ and is also positive in $x_{i_1}, x_{i_2}, \ldots, x_{i_k}$, then

$$F'_{1(x_{i_1}, \ldots, x_{i_k})} \supset F_{1(x_{i_1}, \ldots, x_{i_k})}$$

PROOF: By using Corollary 7.6.2–2, first we form from F_1 the minimum x_{i_1}-function of F_1, $F_{1(x_{i_1})}$. By Theorem 7.6.2,

$$F'_{1(x_{i_1}, \ldots, x_{i_k})} \supset F_{1(x_{i_1})}$$

Then from $F_{1(x_{i_1})}$ we form the minimum x_{i_2}-function of $F_{1(x_{i_1})}$, $F_{1(x_{i_1}, x_{i_2})}$. By Theorem 7.6.2,

$$F'_{1(x_{i_1}, x_{i_2}, \ldots, x_{i_k})} \supset F_{1(x_{i_1}, x_{i_2})}$$

Thus we can form the minimum x_i-functions successively until at last from $F_{1(x_{i_1}, \ldots, x_{i_{k-1}})}$ we form the minimum x_{i_k}-function of $F_{1(x_{i_1}, \ldots, x_{i_{k-1}})}$, $F_{1(x_{i_1}, \ldots, x_{i_k})}$.

By Theorem 7.6.2,

$$F'_{1(x_{i_1}, \ldots, x_{i_k})} \supset F_{1(x_{i_1}, \ldots, x_{i_k})}$$

Q.E.D.

COROLLARY 7.6.3–1. For F_1,

$$F_1 \subset F_{1(x_{i_1})} \subset F_{1(x_{i_1}, x_{i_2})} \subset \ldots \subset F_{1(x_{i_1}, x_{i_2}, \ldots, x_{i_k})}$$

COROLLARY 7.6.3–2. For F_1, $F_{1(x_{i_1}, \ldots, x_{i_k})}$ is unique.

THEOREM 7.6.4. For F_1, $F_{1(x_{i_1}, \ldots, x_{i_k})}$ is independent of the order of the variable, i.e. $F_{1(x_{i_1}, \ldots, x_{i_k})}$ is invariant under all permutations of variables in $(x_{i_1}, \ldots, x_{i_k})$.

PROOF: Consider two arbitrary variables x_{i_1} and x_{i_2} in the set of variables $\{x_{i_1}, x_{i_k}\}$. We put F_1 in the following normal form.

$$F_1 = x_{i_1}p_1 + \bar{x}_{i_1}p_2 + x_{i_2}p_3 + \bar{x}_{i_2}p_4 + x_{i_1}x_{i_2}p_5 + x_{i_1}\bar{x}_{i_2}p_6 + \bar{x}_{i_1}x_{i_2}p_7 + \bar{x}_{i_1}\bar{x}_{i_2}p_8p_9 \qquad (7.6\text{–}8)$$

where $P_1, \ldots, P_9$ each is either 0 or a function independent of x_{i_1} and x_{i_2}.

Then by using Corollary 7.6.2–2, we form $F_{1(x_{i_1})}$, $F_{1(x_{i_1},x_{i_2})}$, $F_{1(x_{i_2})}$, and $F_{1(x_{i_2},x_{i_1})}$ from Equation (7.6–8), as follows.

$$\begin{aligned} F_{1(x_{i_1})} &= x_{i_1}p_1+p_2+x_{i_2}p_3+\bar{x}_{i_2}p_4+x_{i_1}x_{i_2}p_5+x_{i_1}\bar{x}_{i_2}p_6+x_{i_2}p_7+\bar{x}_{i_2}p_8+p_9 \\ &= x_{i_1}p_1+x_{i_2}(p_3+p_7)+\bar{x}_{i_2}(p_4+p_8)+x_{i_1}x_{i_2}p_5+x_{i_1}\bar{x}_{i_2}p_6+(p_2+p_9) \end{aligned} \tag{7.6–9}$$

$$F_{1(x_{i_1},x_{i_2})} = x_{i_1}(p_1+p_6)+x_{i_2}(p_3+p_7)+x_{i_1}x_{i_2}p_5+(p_2+p_4+p_8+p_9) \tag{7.6–10}$$

$$\begin{aligned} F_{1(x_{i_2})} &= x_{i_1}p_1+x_{i_1}p_2+x_{i_2}p_3+p_4+x_{i_1}x_{i_2}p_5+x_{i_1}p_6+x_{i_1}x_{i_2}p_7+x_{i_1}p_8+p_9 \\ &= x_{i_1}(p_1+p_6)+x_{i_1}(p_2+p_8)+x_{i_2}p_3+x_{i_1}x_{i_2}p_5+x_{i_1}x_{i_2}p_7+p_4+p_9 \end{aligned} \tag{7.6–11}$$

$$F_{1(x_{i_2},x_{i_1})} = x_{i_1}(p_1+p_6)+x_{i_2}(p_3+p_7)+x_{i_1}x_{i_2}p_5+(p_2+p_4+p_8+p_9) \tag{7.6–12}$$

From Equations (7.6–10) and (7.6–12), we have

$$F_{1(x_{i_1},x_{i_2})} = F_{1(x_{i_2},x_{i_1})} \tag{7.6–13}$$

It is seen that the positivizing of the function F_1 with respect to a variable x_i is a transformation, and the permutation of variables in the sequence $(x_{i_1}, x_{i_2}, \ldots, x_{i_k})$ is the same as the permutations of the positivizing transformations. Equation (7.6–13) shows that $F_{1(x_{i_1},x_{i_2})}$ is invariant under the transposition of two adjacent variables in any sequence, or the transposition of any two successive transformations.

According to permutation theory, we know that for a fixed finite set of elements, the elements of any finite permutation may be obtained from a given permutation by a finite sequence of transposition of adjacent elements. This implies that $F_{1(x_{i_1},\ldots,x_{i_k})}$ remains invariant when there is any permutation in the variables in the sequence $(x_{i_1}, \ldots, x_{i_k})$. Q.E.D.

$F_{1(x_{i_1},\ldots,x_{i_k})}$ is called the *minimum* $x_{i_1}, \ldots, x_{i_k}$*-function* of F_1.

COROLLARY 7.6.4–1. The minimum $x_{i_1}, \ldots, x_{i_k}$-function of F_1 can be formed from F_1 by deleting all $\bar{x}_{i_1}, \ldots, \bar{x}_{i_k}$'s in the normal disjunctive form of F_1.

COROLLARY 7.6.4–2. The minimum $x_{i_1}, \ldots, x_{i_k}$-function of F_1 can be formed from F_1 by deleting the variables $\bar{x}_{i_1}, \ldots, \bar{x}_{i_k}$ in the normal disjunctive form of F_1, one at a time in any order, by successively forming k minimum functions, the last function formed being the minimum $x_{i_1}, \ldots, x_{i_k}$-function.

THEOREM 7.6.5. For F_1 and F_0, if $F_{1(x_{i_1})}F_{0(\bar{x}_{i_1})} \neq 0$, then F_1 and F_0 are not linearly separable by a threshold function F which is positive in x_{i_1}.

PROOF: Suppose F_1 and F_0 are linearly separable by a threshold function F positive in x_i. Then, $\bar{F}$ is positive in $\bar{x}_i$. By Theorem 7.6.2,

$$F \supset F_{1(x_i)}$$

$$\bar{F} \supset F_{0(\bar{x}_i)}$$

Now

$$F\bar{F} = 0$$

Therefore

$$F_{1(x_i)}F_{0(\bar{x}_i)} = 0$$

This contradicts the assumption. Hence F does not exist. Q.E.D.

THEOREM 7.6.6. If there exist functions $F_{1(x_i)}$ and $F_{0(\bar{x}_i)}$ such that

$$F_{1(x_i)}F_{0(\bar{x}_i)} = 0,$$

and if there also exist functions $F_{1(\bar{x}_i)}$ and $F_{0(x_i)}$, such that

$$F_{1(\bar{x}_i)}F_{0(x_i)} = 0,$$

then there exist functions F_1' and F_0', both independent of x_i, such that

$$F_1' \supset F_1, \quad F_0' \supset F_0, \quad \text{and} \quad F_1'F_0' = 0$$

PROOF: Let

$$F_1 = x_iP + \bar{x}_iQ + R$$

$$F_0 = x_iS + \bar{x}_iT + U$$

where P, Q, R, S, T, and U are each either 0 or a function independent of x_i. Then by Theorem 7.6.2 and Corollary 7.6.2–2, we obtain

$$F_{1(x_i)} = x_iP + Q + R \tag{7.6–14}$$

$$F_{0(\bar{x}_i)} = S + \bar{x}_iT + U \tag{7.6–15}$$

According to the hypothesis, we have

$$F_{1(x_i)}F_{0(\bar{x}_i)} = x_iP(S+U) + \bar{x}_iT(Q+R) + x_i\bar{x}_iPT + (Q+R)(S+U) = 0$$

Therefore

$$PS = PU = QS = QT = QU = RS = RT = RU = 0 \tag{7.6–16}$$

Similarly

$$F_{1(\bar{x}_i)} = P + \bar{x}_iQ + R \tag{7.6–17}$$

$$F_{0(x_i)} = x_i S + T + U \tag{7.6–18}$$

and

$$F_{1(\bar{x}_i)} F_{0(x_i)} = \bar{x}_i Q(T+U) + x_i S(P+R) + \bar{x}_i x_i QS + (P+R)(T+U) = 0$$

Therefore

$$PS = PT = PU = QT = QU = RS = RT = RU = 0 \tag{7.6–19}$$

Set

$$F'_1 = P+Q+R \supset F_1 \tag{7.6–20}$$

$$F'_0 = S+T+U \supset F_0 \tag{7.6–21}$$

By Equations (7.6–16) and (7.6–19) we have

$$F'_1 F'_0 = 0$$

Q.E.D.

The functions F'_1 and F'_0 obtained in Theorem 7.6.6 are called *minimum x_i^*-functions* of F_1 and F_0, respectively.

It is obvious that minimum x_i^*-functions are positive both in x_i and $\bar{x}_i$, i.e. independent of x_i.

THEOREM 7.6.7. If there exist functions $F_{1(x_i)}$, $F_{0(\bar{x}_i)}$, and $F_{1(\bar{x}_i)}$, $F_{0(x_i)}$, such that

$$F_{1(x_i)} F_{0(\bar{x}_i)} = 0$$

$$F_{1(\bar{x}_i)} F_{0(x_i)} = 0$$

and if any one set of them, say $\{F_{1(x_i)}, F_{0(\bar{x}_i)}\}$, is linearly separable, then,

(1) The other set, $\{F_{1(\bar{x}_i)}, F_{0(x_i)}\}$, is not necessarily linearly separable:

(2) The minimum x_i^*-functions F'_1 and F'_0 are not necessarily linearly separable.

PROOF: We shall prove this theorem by using a counter-example.

(1) In the example given in Section 7.8, the set $\{F_{1(x_1, x_2, x_3, \bar{x}_4, x_5, x_6)}, F_{0(\bar{x}_1, \bar{x}_2, \bar{x}_3, x_4, \bar{x}_5, \bar{x}_6)}\}$ is not linearly separable.

(2) In the same example, the set $\{F'_{1(x_1, x_2, x_3, x_5, x_6)}, F'_{0(\bar{x}_1, \bar{x}_2, \bar{x}_3, \bar{x}_5, \bar{x}_6)}\}$ is also not linearly separable. Q.E.D.

Now we have a systematic method for finding the unate functions F_{1U} and F_{0U}. Suppose we start from x_i. From F_1 we can form $F_{1(x_i)}$ positive in x_i and $F_{1(\bar{x}_i)}$ positive in $\bar{x}_i$. Similarly from F_0 we can form $F_{0(\bar{x}_i)}$ positive in $\bar{x}_i$ and $F_{0(x_i)}$ positive in x_i. Then we have four possible cases.

Case (1).

$$F_{1(x_i)} F_{0(\bar{x}_i)} \neq 0, \quad F_{1(\bar{x}_i)} F_{0(x_i)} \neq 0$$

By Theorem 7.6.5, F_1 is not 1-realizable by a threshold function either positive in x_i or positive in $\bar{x}_i$. So F_1 is not 1-realizable.

Case (2).

$$F_{1(x_i)}F_{0(\bar{x}_i)} = 0, \quad F_{1(\bar{x}_i)}F_{0(x_i)} \neq 0.$$

F_1 is not 1-realizable by a threshold function positive in $\bar{x}_i$, but may be 1-realizable by a threshold function positive in x_i. So we must change F_1 into $F_{1(x_i)}$ and F_0 into $F_{0(x_i)}$. Then proceed to the positivizing with respect to the next variable.

Case (3).

$$F_{1(x_i)}F_{0(\bar{x}_i)} \neq 0, \quad F_{1(\bar{x}_i)}F_{0(x_i)} = 0.$$

F_1 is not 1-realizable by a threshold function positive in x_i, but may be 1-realizable by a threshold function positive in $\bar{x}_i$. So we must change F_1 into $F_{1(\bar{x}_i)}$ and F_0 into $F_{0\,(x_i)}$. Then proceed to the positivizing with respect to the next variable.

Case (4).

$$F_{1(x_i)}F_{0(\bar{x}_i)} = 0, \quad F_{1(\bar{x}_i)}F_{0(x_i)} = 0.$$

F_1 may be 1-realizable by a threshold function positive in x_i or in $\bar{x}_i$. In this case we have to try both ways in order to exhaust all the possibilities. However, we can leave x_i for the moment and proceed to the next variable. After the positivizing with respect to other variables are made, we can come back to x_i again, and at that time we might find that for x_i, $F'_{1(x_i)}$ $F'_{0(\bar{x}_i)} \neq 0$, or/and $F_{1(\bar{x}_i)}F_{0(x_i)} \neq 0$, because as the function is made positive in more variables, more vertices are added to F_1 and F_0, and there will be more chance for $F_{1(x_i)}$ and $F_{0(\bar{x}_i)}$ or/and $F_{1(\bar{x}_i)}$ and $F_{0(x_i)}$ to have common vertices.

When there are no don't care vertices, $F_1 = F$ and $F_0 = \bar{F}$. If F is positive in x_i, F cannot be positive in $\bar{x}_i$ unless F is independent of x_i. When the number of don't care vertices is small, the space of don't care vertices is small too, and it does not happen very often that F_1 can be positive both in x_i and also in $\bar{x}_i$. When the number of don't care vertices is large and the F_1 vertices and the F_0 vertices are "far apart", 1-realization becomes a relatively easy thing. Therefore Case (4) mentioned above is actually not as serious as it first appears to be.

To check if $F_{1(x_i)}F_{0(\bar{x}_i)} = 0$, we can check the product of every prime implicant of $F_{1(x_i)}$ and that of $F_{0(\bar{x}_i)}$ to see if it is zero or not. The product is zero, iff it subsumes both the unnegated and the negated form of some variable x_i.

Once F_1 and F_0 are changed into unate functions F_{1U} and F_{0U}, respectively, the next problem is whether it is necessary or desirable to find the primary ordering of weights from F_{1U} and F_{0U} or not. The answer is no.

In ordinary 1-realization, complete primary ordering is a necessary condition, as it is equivalent to 2-monotonicity. Since primary ordering can be found from the function directly, without the necessity of resorting to the set of inequalities, it serves as a simple and convenient preliminary step. Besides, if the given function is a threshold function, primary ordering will help to reduce the number of inequalities to $p'+q'$.

But in the case of 1-realization with don't care vertices, complete primary ordering is not a necessary condition. If we cannot obtain a complete primary ordering from the function F_{1U}, F_1 may still be 1-realizable. So we save this step and leave the ordering of weights entirely to the inequalities. In fact, the successive higher ordering method is particularly suitable for this kind of comparison work, because it will reveal every partial ordering, whether primary or higher, as long as it exists, even if there is no complete primary ordering. So once F_{1U} and F_{0U} are found, we simply consider the problem as an ordinary 1-realization problem and write down the inequalities, except that in this case while the set of minimal vertices $\{L\}$ are still taken from F_{1U}, the set of maximal vertices $\{H\}$ are taken from F_{0U} instead of from $\bar{F}_{1U}$.

7.7 PROCEDURE OF TESTING OF FUNCTIONS WITH DON'T CARE VERTICES

Now we can formulate a procedure as follows:

(1) Express the given functions F_1 and F_0 in their irredundant normal disjunctive form.

(2) Using the method described in Section 7.6, form a pair or pairs of unate functions F_{1U} and F_{0U} by the successive positivizing of F_1 and F_0, such that for each pair

(a) $F_{1U} \supset F_1, \quad F_{0U} \supset F_0$

(b) $F_{1U} \; F_{0U} = 0$

(c) If F_{1U} is positive in a variable x_i, then F_{0U} is positive in $\bar{x}_i$, and vice versa.

If no such pairs of unate functions exist, F_1 is not 1-realizable. If there is one or more such pairs, express them in irredundant normal disjunctive form.

(3) For each pair of unate functions $\{F_{1U}, F_{0U}\}$, consider the set of minimal vertices, $\{L\}$, of F_{1U} and the set of maximal vertices, $\{H\}$, of F_{0U} just as those sets $\{L\}$ and $\{H\}$ of F and $\bar{F}$, respectively, in the case of ordinary 1-realization without don't care vertices. Write down the inequalities corresponding to $\{L\}$ and $\{H\}$ and proceed according to the

successive higher ordering method, except that those steps related to 2-monotonicity and primary ordering are omitted. If for all pairs of unate functions some contradiction is revealed, then F_1 is not 1-realizable. If for one or more pairs of unate functions there is no contradiction, then an assignment of weights can be made, a threshold function F can be found, and F_1 is 1-realizable. The threshold function F found for any pair of unate functions 1-realizes the given function F_1.

7.8 EXAMPLE OF TESTING AND REALIZATION OF A FUNCTION WITH DON'T CARE VERTICES

$$F_1 = x_1x_2\bar{x}_3+x_1x_3x_4\bar{x}_5+x_1\bar{x}_2x_3x_5+x_1x_3x_6+\bar{x}_1x_2x_3x_5 \\ +\bar{x}_1x_2x_3x_6+x_3x_4x_5x_6$$

$$F_0 = x_1\bar{x}_2\bar{x}_3\bar{x}_5+\bar{x}_2\bar{x}_3\bar{x}_4+\bar{x}_1\bar{x}_4\bar{x}_5\bar{x}_6+\bar{x}_1x_2\bar{x}_3\bar{x}_5+\bar{x}_1\bar{x}_3\bar{x}_4x_6 \\ +\bar{x}_1\bar{x}_2x_4\bar{x}_6+\bar{x}_1\bar{x}_2x_3\bar{x}_5$$

(1) F_1 and F_0 are already in irredundant normal disjunctive form.

(2) By inspection, we have

(a) $F_{1(x_1)}F_{0(\bar{x}_1)} = 0$

$F_{1(\bar{x}_1)}F_{0(x_1)} \neq 0$

Take $F_{1(x_1)}$ and $F_{0(\bar{x}_1)}$

$$F_{1(x_1)} = x_1x_2\bar{x}_3+x_1x_3x_4\bar{x}_5+x_1\bar{x}_2x_3x_5+x_1x_3x_6 \\ +x_2x_3x_5+x_2x_3x_6+x_3x_4x_5x_6$$

$$F_{0(x_1)} = \bar{x}_2\bar{x}_3\bar{x}_5+\bar{x}_2\bar{x}_3\bar{x}_4+\bar{x}_1\bar{x}_4\bar{x}_5\bar{x}_6+\bar{x}_1x_2\bar{x}_3\bar{x}_5 \\ \bar{x}_1\bar{x}_3\bar{x}_4x_6+\bar{x}_1\bar{x}_2x_4\bar{x}_6+\bar{x}_1\bar{x}_2x_3\bar{x}_5$$

(b) $F_{1(x_1x_2)}F_{0(\bar{x}_1\bar{x}_2)} = 0$

$F_{1(x_1\bar{x}_2)}F_{0(\bar{x}_1x_2)} \neq 0$

Take $F_{1(x_1x_2)}$ and $F_{0(\bar{x}_1\bar{x}_2)}$

$$F_{1(x_1x_2)} = x_1x_2\bar{x}_3+x_1x_3x_4\bar{x}_5+x_1x_3x_5+x_1x_3x_6 \\ +x_2x_3x_5+x_2x_3x_6+x_3x_4x_5x_6$$

$$F_{0(\bar{x}_1\bar{x}_2)} = \bar{x}_2\bar{x}_3\bar{x}_5+\bar{x}_2\bar{x}_3\bar{x}_4+\bar{x}_1\bar{x}_4\bar{x}_5\bar{x}_6+\bar{x}_1\bar{x}_3\bar{x}_5 \\ +\bar{x}_1\bar{x}_3\bar{x}_4x_6+\bar{x}_1\bar{x}_2x_4\bar{x}_6+\bar{x}_1\bar{x}_2x_3\bar{x}_5$$

(c) $F_{1(x_1x_2x_3)}F_{0(\bar{x}_1\bar{x}_2\bar{x}_3)} = 0$

$F_{1(x_1x_2\bar{x}_3)}F_{0(\bar{x}_1\bar{x}_2x_3)} \neq 0$

Take $F_{1(x_1x_2x_3)}$ and $F_{0(\bar{x}_1\bar{x}_2\bar{x}_3)}$

$$F_{1(x_1x_2x_3)} = x_1x_2+x_1x_3x_4\bar{x}_5+x_1x_3x_5+x_1x_3x_6 +x_2x_3x_5+x_2x_3x_6+x_3x_4x_5x_6$$

$$F_{0(\bar{x}_1\bar{x}_2\bar{x}_3)} = \bar{x}_2\bar{x}_3\bar{x}_5+\bar{x}_2\bar{x}_3\bar{x}_4+\bar{x}_1\bar{x}_4\bar{x}_5\bar{x}_6+\bar{x}_1\bar{x}_3\bar{x}_5 +\bar{x}_1\bar{x}_3\bar{x}_4x_6+\bar{x}_1\bar{x}_2x_4\bar{x}_6+\bar{x}_1\bar{x}_3\bar{x}_5$$

(d) $F_{1(x_1x_2x_3x_4)}F_{0(\bar{x}_1\bar{x}_2\bar{x}_3\bar{x}_4)} = 0$

$F_{1(x_1x_2x_3\bar{x}_4)}F_{0(\bar{x}_1\bar{x}_2\bar{x}_3x_4)} = 0$

Since there are two choices, we leave positivizing with respect to x_4 to the end, and proceed to the positivizing with respect to the next variable x_5.

(e) $F_{1(x_1x_2x_3x_4)}F_{0(\bar{x}_1\bar{x}_2\bar{x}_3\bar{x}_4)} = 0$

$F_{1(x_1x_2x_3\bar{x}_5)}F_{0(\bar{x}_1\bar{x}_2\bar{x}_3x_5)} \neq 0$

Take $F_{1(x_1x_2x_3x_5)}$ and $F_{0(\bar{x}_1\bar{x}_2\bar{x}_3\bar{x}_5)}$

$$F_{1(x_1x_2x_3x_5)} = x_1x_2+x_1x_3x_4+x_1x_3x_5+x_1x_3x_6 +x_2x_3x_5+x_2x_3x_6+x_3x_4x_5x_6$$

$$F_{0(\bar{x}_1\bar{x}_2\bar{x}_3\bar{x}_5)} = \bar{x}_2\bar{x}_3\bar{x}_5+\bar{x}_2\bar{x}_3\bar{x}_4+\bar{x}_1\bar{x}_4\bar{x}_5\bar{x}_6+\bar{x}_1\bar{x}_3\bar{x}_5 +\bar{x}_1\bar{x}_3\bar{x}_4x_6+\bar{x}_1\bar{x}_2x_4\bar{x}_6+\bar{x}_1\bar{x}_2\bar{x}_5$$

(f) $F_{1(x_1x_2x_3x_5x_6)}F_{0(\bar{x}_1\bar{x}_2\bar{x}_3\bar{x}_5\bar{x}_6)} = 0$

$F_{0(x_1x_2x_3x_5\bar{x}_6)}F_{0(\bar{x}_1\bar{x}_2\bar{x}_3\bar{x}_5x_6)} \neq 0$

Take $F_{1(x_1x_2x_3x_5x_6)}$ and $F_{0(\bar{x}_1\bar{x}_2\bar{x}_3\bar{x}_5\bar{x}_6)}$

$$F_{1(x_1x_2x_3x_5x_6)} = x_1x_2+x_1x_3x_4+x_1x_3x_5+x_1x_3x_6 +x_2x_3x_5+x_2x_3x_6+x_3x_4x_5x_6$$

$$F_{0(\bar{x}_1\bar{x}_2\bar{x}_3\bar{x}_5\bar{x}_6)} = \bar{x}_2\bar{x}_3\bar{x}_5+\bar{x}_2\bar{x}_3\bar{x}_4+\bar{x}_1\bar{x}_4\bar{x}_5\bar{x}_6+\bar{x}_1\bar{x}_3\bar{x}_5 +\bar{x}_1\bar{x}_3\bar{x}_4+\bar{x}_1\bar{x}_2x_4\bar{x}_6+\bar{x}_1\bar{x}_2\bar{x}_5$$

(g) After the positivizing with respect to the last variable x_4 the functions obtained will be unate functions.

$$F_{1(x_1x_2x_3x_4x_5x_6)}F_{0(\bar{x}_1\bar{x}_2\bar{x}_3\bar{x}_4\bar{x}_5\bar{x}_6)} = 0$$

$$F_{1(x_1x_2x_3\bar{x}_4x_5x_6)}F_{0(\bar{x}_1\bar{x}_2\bar{x}_3x_4\bar{x}_5\bar{x}_6)} = 0$$

By Theorem 7.6.6, there exist functions $F'_{1(x_1x_2x_3x_5x_6)}$ and $F'_{0(\bar{x}_1\bar{x}_2\bar{x}_3\bar{x}_5\bar{x}_6)}$ independent of x_4 such that

$$F'_{1(x_1x_2x_3x_4x_6)}\ F'_{0(\bar{x}_1\bar{x}_2\bar{x}_3\bar{x}_5\bar{x}_6)} = 0$$

Now there are three pairs of unate functions:

$$\text{PAIR } 1: F_{1U_1} = F_{1(x_1x_2x_3x_4x_5x_6)}, \quad F_{0U_1} = F_{0(\bar{x}_1\bar{x}_2\bar{x}_3\bar{x}_4\bar{x}_5\bar{x}_6)}$$

$$\text{PAIR } 2: F_{1U_2} = F_{1(x_1x_2x_3\bar{x}_4x_5x_6)}, \quad F_{0U_2} = F_{0(\bar{x}_1\bar{x}_2\bar{x}_3x_4\bar{x}_5\bar{x}_6)}$$

$$\text{PAIR } 3: F_{1U_3} = F'_{1(x_1x_2x_3x_5x_6)}, \quad F_{0U_3} = F'_{0(\bar{x}_1\bar{x}_2\bar{x}_3\bar{x}_5\bar{x}_6)}$$

Therefore we have to test all the three pairs for 1-realizability by using the successive higher ordering method.

(3) In order to show the application of this method clearly, we follow the procedure in Section 7.3, except that those steps related to 2-monotonicity and primary ordering are omitted.

PAIR 1:

$$F_{1U_1} = x_1x_2+x_1x_3x_4+x_1x_3x_5+x_1x_3x_6+x_2x_3x_5+x_2x_3x_6 +x_3x_4x_5x_6$$

$$F_{0U_1} = \bar{x}_2\bar{x}_3\bar{x}_5+\bar{x}_2\bar{x}_3\bar{x}_4+\bar{x}_1\bar{x}_4\bar{x}_5\bar{x}_6+\bar{x}_1\bar{x}_3\bar{x}_5+\bar{x}_1\bar{x}_3\bar{x}_4\pm\bar{x}_1\bar{x}_2\bar{x}_6 +\bar{x}_1\bar{x}_2\bar{x}_5$$

Step 1: F_{1U_1} is a unate function.

Step 2: F_{0U_1} is a unate function.

Step 3: F_{1U_1} and F_{0U_1} are already in irredundant normal disjunctive form.

Step 4: Primary ordering is not possible, nor is it required.

Step 5: The inequalities are put in tabular form as shown in Table XI.

TABLE XI

	j	a_1	a_2	a_3	a_4	a_5	a_6	S_{Lj}		k	a_1	a_2	a_3	a_4	a_5	a_6	S_{Rk}
	1	1	1					2		1	1			1		1	3
	2	1		1	1			3		2	1				1	1	3
	3	1		1		1		3		3		1	1				2
g_{Lj}	4	1		1			1	3	$> g_{Rk}$	4		1		1		1	3
	5		1	1		1		3		5		1			1	1	3
	6		1	1			1	3		6			1	1	1		3
	7			1	1	1	1	4		7			1	1		1	3

Steps 6 and 7: These steps are omitted because there is no primary ordering.

Step 8: Condition *b* is satisfied. So we proceed to Step 9.

Step 9: Condition *b* is satisfied. So we proceed to Step 10.

Step 10: From Table XI, we find the following irredundant explicit relations.

$g_{L1}-g_{R1} \qquad a_2 > a_4+a_6$
$g_{L1}-g_{R2} \qquad a_2 > a_5+a_6$
$g_{L1}-g_{R3} \qquad a_1 > a_3$
$g_{L1}-g_{R4} \qquad a_1 > a_4+a_6$
$g_{L1}-g_{R5} \qquad a_1 > a_5+a_6$
$g_{L2}-g_{R1} \qquad a_3 > a_6$
$g_{L4}-g_{R1} \qquad a_3 > a_4$
$g_{L4}-g_{R2} \qquad a_3 > a_5$

Condition *a* is satisfied. Out of these 8 relations we can take three only. Set

$$\begin{aligned} a_2 &= \Delta a_2+a_5+a_6 \\ a_1 &= \Delta a_1+a_5+a_6 \\ a_3 &= \Delta a_3+a_6 \end{aligned} \qquad (7.8\text{–}1)$$

Substituting Equation (7.8–1) back into Table XI and eliminating the common term $C = a_6$, we have Table XII.

TABLE XII

	j	Δa_1	Δa_2	Δa_3	a_4	a_5	a_6	S_{Lj}		k	Δa_1	Δa_2	Δa_3	a_4	a_5	a_6	S_{Rk}
	1	1	1			2	1	5		1	1			1	1	1	4
	2	1		1	1	1	1	5		2	1				2	1	4
	3	1		1		2	1	5		3		1	1		1	1	4
g_{Lj}	4	1		1		1	2	5	$> g_{Rk}$	4		1		1	1	1	4
	5		1	1		2	1	5		5		1			2	1	4
	6		1	1		1	2	5		6			1	1	1		3
	7			1	1	1	1	4		7			1	1		1	3

Step 8: Condition *b* is satisfied. So we proceed to Step 9.

Step 9: Condition *b* is satisfied. So we proceed to Step 10.

Step 10: From Table XII, we find

$g_{L7}-g_{R3} \qquad a_4 > \Delta a_2$
$g_{L7}-g_{R4} \qquad \Delta a_3 > \Delta a_2$

Condition *a* is satisfied. Set

$$\begin{aligned} a_4 &= \Delta a_4+\Delta a_2 \\ \Delta a_3 &= \Delta^2 a_3+\Delta a_2 \end{aligned} \qquad (7.8\text{–}2)$$

Substituting Equation (7.8–2) back into Table XII, we obtain Table XIII.

TABLE XIII

	j	Δa_1	Δa_2	$\Delta^2 a_3$	Δa_4	a_5	a_6	S_{Lj}		k	Δa_1	Δa_2	$\Delta^2 a_3$	Δa_4	a_5	a_6	S_{Rk}
	1	1	1			2	1	5		1	1	1		1	1	1	5
	2	1	2	1	1	1	1	7		2	1				2	1	4
	3	1	1	1		2	1	6		3		2	1		1	1	5
g_{Lj}	4	1	1	1		1	2	6	$> g_{Rk}$	4		2		1	1	1	5
	5		2	1		2	1	6		5		1			2	1	4
	6		2	1		1	2	6		6		2	1	1	1		5
	7		2	1	1	1	1	6		7		2	1	1		1	5

Step 8: Condition b is satisfied. So we proceed to Step 9.

Step 9: Condition b is satisfied. So we proceed to Step 10.

Step 10: From Table XIII, we find

$$g_{L1} - g_{R1} \qquad a_5 > \Delta a_4$$

Condition a is satisfied. Set

$$a_5 = \Delta a_5 + \Delta a_4 \tag{7.8–3}$$

Substituting Equation (7.8–3) back into Table XIII and eliminating the new common term $C = \Delta a_4$, we obtain Table XIV.

TABLE XIV

	j	Δa_1	Δa_2	$\Delta^2 a_3$	Δa_4	Δa_5	a_6	S_{Lj}		k	Δa_1	Δa_2	$\Delta^2 a_3$	Δa_4	Δa_5	a_6	S_{Rk}
	1	1	1		1	2	1	6		1	1	1		1	1	1	5
	2	1	2	1	1	1	1	7		2	1			1	2	1	5
	3	1	1	1	1	2	1	7		3		2	1		1	1	5
g_{Lj}	4	1	1	1		1	2	6	$> g_{Rk}$	4		2		1	1	1	5
	5		2	1	1	2	1	7		5		1		1	2	1	5
	6		2	1		1	2	6		6		2	1	1	1		5
	7		2	1	1	1	1	6		7		2	1			1	4

Step 8: Condition a is satisfied. So we proceed to Step 11.

Step 11: Assign

$$\Delta a_1 = \Delta a_2 = \Delta^2 a_3 = \Delta a_4 = \Delta a_5 = a_6 = 1$$

Substituting these values into Equations (7.8–1), (7.8–2), and (7.8–3) we obtain

$$a_1 = 4 \qquad a_2 = 4 \qquad a_3 = 3$$
$$a_4 = 2 \qquad a_5 = 2 \qquad a_6 = 1$$

And from Table XI, we have

$$T = 8$$

Therefore

$$R[F] = (4, 4, 3, 2, 2, 1; 8)$$
$$F = x_1x_2 + x_1x_3x_4 + x_1x_3x_5 + x_1x_3x_6 + x_1x_4x_5 + x_2x_3x_4 + x_2x_3x_5 + x_2x_3x_6 + x_2x_4x_5 + x_3x_4x_5x_6$$

PAIR 2:

$$F_{1U_2} = x_1x_2 + x_1x_3 + x_2x_3x_5 + x_2x_3x_6 + x_3x_5x_6$$
$$F_{0U_2} = \bar{x}_2\bar{x}_3 + \bar{x}_1\bar{x}_5\bar{x}_6 + \bar{x}_1\bar{x}_3 + \bar{x}_1\bar{x}_2x_4\bar{x}_6 + \bar{x}_1\bar{x}_2\bar{x}_5$$

Using the successive higher ordering method, as for Pair 1, we find that F_{1U_2} and F_{0U_2} are not linearly separable.

PAIR 3:

$$F_{1U_3} = x_1x_2 + x_1x_3 + x_2x_3x_5 + x_2x_3x_6 + x_3x_5x_6$$
$$F_{0U_3} = \bar{x}_2\bar{x}_3 + \bar{x}_1\bar{x}_5\bar{x}_6 + \bar{x}_1\bar{x}_3 + \bar{x}_1\bar{x}_2\bar{x}_6 + \bar{x}_1\bar{x}_2\bar{x}_5$$

Similarly we find that F_{1U_3} and F_{0U_3} are not linearly separable.

REFERENCES

1. Coates, C. L., and Lewis, P. M. II. (1963). Threshold gate realizations of logical functions with don't cares, *Proc. 4th Ann. Symp. Switching Circuit Theory and Logical Design*, 41–52.
2. Hu, S. T. (1965). "Threshold logic", University of California Press, U.S.A.
3. Sheng, C. L. (1964). A method for testing and realization of threshold functions, *I.E.E.E. Trans. electronic Comput.* EC**13,** 232–239.
4. Sheng, C. L., and Hwa, H. R. (1966). The successive higher ordering method for the testing and realization of threshold functions, *I.E.E.E. Trans. electronic Comput.* EC**15,** 212–220.
5. Sheng, C. L., and Hwa, H. R. (1966). Testing and realization of threshold functions with don't cares, Technical Report No. 66–5, Department of Electrical Engineering, University of Ottawa.
6. Winder, R. O. (1962). "Threshold logic", Doctoral dissertation, Mathematics Department, Princeton University, U.S.A.

CHAPTER 8

Compound Synthesis

8.1 INTRODUCTION

Since the number of threshold functions of n variables is only a small fraction of the total number of switching functions of n variables, and since the fraction decreases rapidly with the increase of n, an arbitrary switching function is in general not 1-realizable, and needs several threshold logic elements for realization. The synthesis of a network of threshold logic elements for the realization of a given switching function is called compound synthesis. In general, compound synthesis is not unique, and a given function can be realized in many different ways. For the sake of economy, a basic criterion for compound synthesis is the number of threshold logic elements required, i.e. it is desirable to synthesize a network composed of a minimum number of threshold logic elements. However, unlike the two-stage minimization of a switching function in terms of AND and OR gates, the minimization of threshold-logic network is much more difficult and has no simple and straightforward solution. Therefore it is not always justified to search for the exact optimum solution, and sometimes we have to accept a good or close-to-optimum solution.

Compound synthesis can be studied in general, with no particular constraint, as was done by Stram,[26] Minnick,[20] Fischler,[8] Winder,[28] Mattson,[17] Cameron,[5] Hopcroft,[10] Hu,[13] Sheng,[25] etc. However, most of the work done in compound synthesis is the study of networks composed of a certain special type of threshold logic element exclusively, for instance a majority gate with a fixed number of inputs, especially the three-input majority gate.[16, 18, 7, 3, 1, 11, 20, 9, 4, 23, 21, 27, 12] Threshold-logic networks can also be studied with practical constraints, such as sensitivity require ments, as was done by Coates and Lewis,[6] Lewis and Coates,[15] etc. Also compound synthesis can be limited to the realization of a restricted class of switching functions, say symmetrical functions, as was done by Minnick,[19] Muroga, Toda, and Takasu,[22] Kautz,[14] Sheng,[24] etc.

In this chapter we shall study in some detail general compound synthesis for the realization of an arbitrary switching function, without considerations of reliability, number of inputs, size of weights, sensitivity requirements, etc. We shall also describe briefly one of the systematic methods for the synthesis of threshold-logic networks composed exclusively of three-input majority gates. Since compound synthesis for the realization of symmetric functions is somewhat restricted, we shall explain its basic concepts only, without giving the details of the specific methods. As to the satisfaction of sensitivity requirements, we consider it as a separate problem which can be detached from the realization problem completely, and has still less to do with compound synthesis. It will be studied in some detail in Chapter 9.

For an arbitrary switching function, it is unavoidable that some variable or variables appear in both the uncomplemented and complemented forms. It is assumed that both uncomplemented and complemented variables are available, and that the weights of the input variables are all positive.

In a network of threshold logic elements, the outputs of some elements are connected as inputs to other elements. We shall call these inputs *function-inputs*, because they themselves are functions of the n variables $\{x_i\}$. So far as the element receiving a function-input is concerned, a function-input is simply a binary variable. Thus if both uncomplemented and complemented forms of the function-inputs are available, the weights of the function-inputs may be assumed to be all positive too. However, to simplify the network we assume that all the function-inputs are in uncomplemented form only. Therefore, in general, the weight of a function-input may be either positive or negative.

8.2 BASIC THEOREMS

We shall study compound synthesis first from the decomposition point of view. In order to realize the function with a relatively small number of threshold logic elements, decomposition will be applied to a general Boolean function along threshold functions instead of variables.

In the following we shall use the symbol F to denote an arbitrary Boolean function, F_T to denote a threshold function, and F_U to denote a unate function.

THEOREM 8.2.1. If a Boolean function F of n variables can be expressed as the sum of m threshold functions, i.e.

$$F = F_{T1}+F_{T2}+\ldots+F_{Tm},$$

then F is realizable with m threshold logic elements, or is m-realizable.

PROOF: First let us consider the case of $m = 2$, i.e. $F = F_{T1}+F_{T2}$. Since F_{T1} and F_{T2} are each 1-realizable, F can be realized as the output of an OR gate with F_{T1} and F_{T2} as the two inputs, as shown in Fig. 5(a).

Now the OR gate can be replaced by a threshold logic element of two

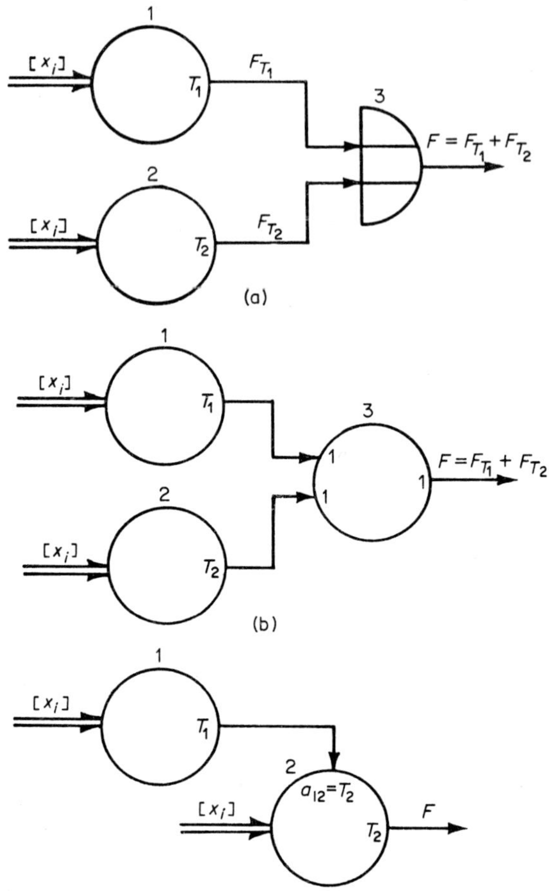

FIG. 5. Realization of $F = F_{T1} + F_{T2}$.

inputs, F_{T1} and F_{T2}, each with a weight of 1, and a threshold value of 1, as shown in Fig. 5(b).

Here the function F is realized with 3 or $m+1$ elements. If the first element is connected to the second element as a function-input with a weight $a_{12} = T_2$, as shown in Fig. 5(c), then the output function is

$$F' = F_{T1} + \overline{F}_{T1} F_{T2} = F_{T1} + F_{T2} = F$$

Thus Theorem 8.2.1 is proved for the case of $m = 2$.

For the case of $m = 3$, we have

$$F = F_{T1} + F_{T2} + F_{T3}$$

We can realize $F_2 = F_{T1} + F_{T2}$ with 2 elements and then connect

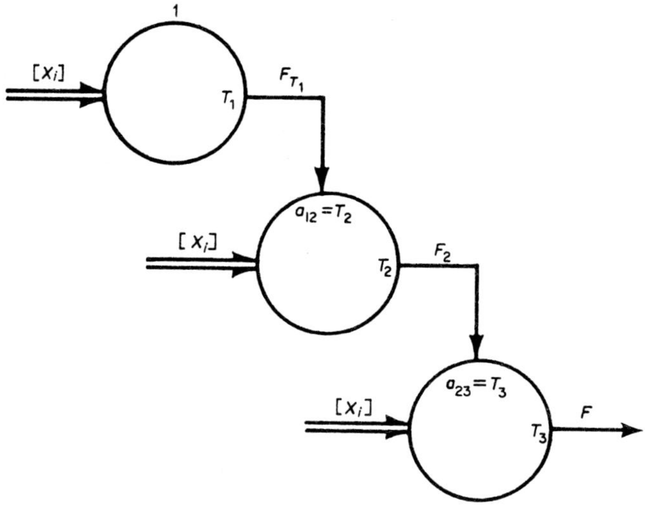

FIG. 6. Realization of $F = F_{T1}+F_{T2}+F_{T3}$.

F_2 as an input to the third element with a weight of T_3, as shown in Fig. 6. Then

$$F' = F_2+\bar{F}_2F_{T3} = F_2+F_{T3} = F_{T1}+F_{T2}+F_{T3} = F$$

It is to be noted that F_2 is in general not a threshold function. If it is, then F_2 becomes 1-realizable and F becomes 2-realizable.

The same reasoning can be extended to the general case of m being equal to any positive integer. Q.E.D.

THEOREM 8.2.2. If a Boolean function F of n variables can be expressed as the product of m threshold functions, i.e.

$$F = F_{T1}F_{T2}\dots F_{Tm},$$

then F is realizable with m threshold logic elements, or is m-realizable.

PROOF: Theorem 8.2.2 is the dual of Theorem 8.2.1. Let us consider the case of $m = 2$ first. F can be realized with two threshold logic elements and an AND gate, as shown in Fig. 7(a). The AND gate may be replaced by a threshold logic element of two inputs, F_{T1} and F_{T2}, each of weight 1, and with a threshold value of 2, as shown in Fig. 7(b). Finally threshold logic element 1 may be connected to threshold logic element 2 as a function-input with a weight $a_{12} > \sum_{i=1}^{n} a_{i(2)}$, where $\{a_{i(2)}\}$ = weights of

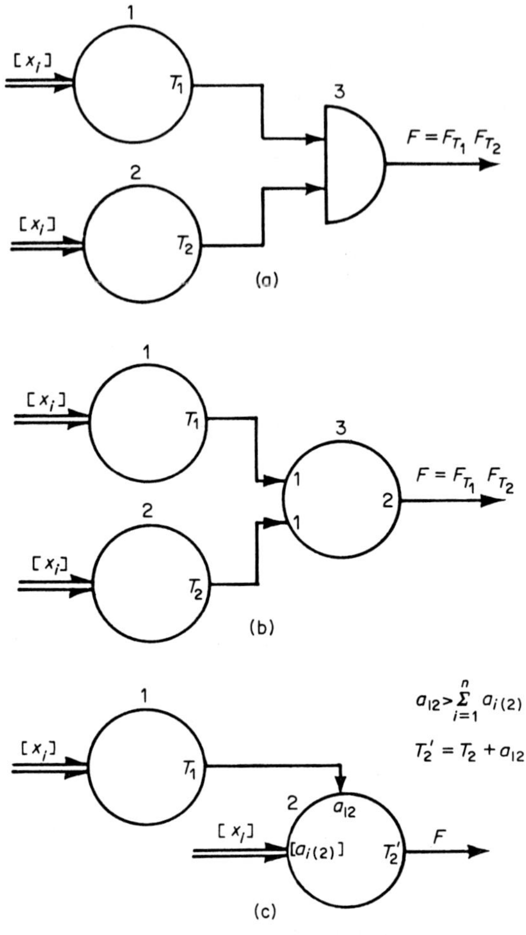

FIG. 7. Realization of $F = F_{T1}F_{T2}$.

the n input variables $\{x_i\}$ of element 2; and by raising the threshold value of clement 2 from T_2 to $T_2' + a_{12}$, the output of element 2 is the function F as shown in Fig. 7(c). The reason for setting $a_{12} > \sum_{i=1}^{n} a_{i(2)}$ is to ensure that there will be no output at 2 for any combination of input values, when F_{T1} is "0".

This reasoning can easily be extended to the general case of m being equal to any positive integer. Q.E.D.

Theorems 8.2.1 and 8.2.2 actually prove the realizability of a function which is equal to the sum (or product) of a non-threshold function and a threshold function by connecting the former to the latter as a function-input. Now we can extend it to the case of realizing a function which is the sum (or product) of two functions both of which are not threshold functions, by connecting one to the other as an input.

THEOREM 8.2.3. (a) If $F = F_1+F_2$, where F, F_1, and F_2 are arbitrary Boolean functions, and if F_1 and F_2 can be realized with j and k threshold logic elements, respectively, then F can be realized with $j+k$ threshold logic elements.

(b) If $F = F_1F_2$, where F, F_1, and F_2 are arbitrary Boolean functions, and if F_1 and F_2 can be realized with j and k threshold logic elements, respectively, then F can be realized with $m = j+k$ threshold logic elements.

PROOF: (a) Let us connect F_1 as a function-input to the last-stage element of F_2 with a weight $a_{12} = T_2$, the original threshold value of the

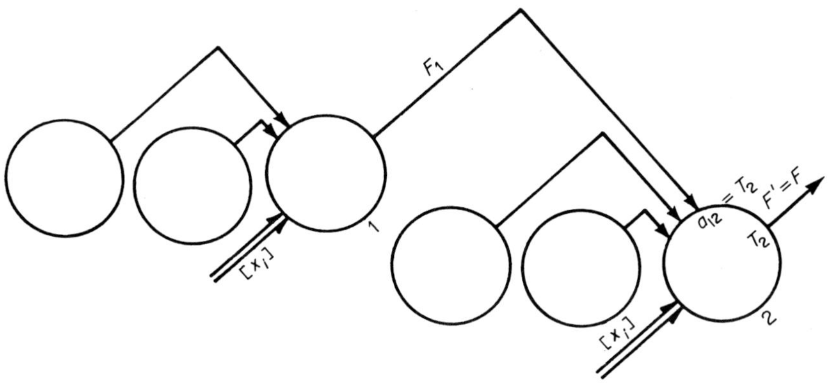

FIG. 8. Realization of $F = F_1+F_2$.

last-stage element realizing F_2, and with T_2 unchanged, as shown in Fig. 8. The truth conditions of the functions are shown in Table XV.

TABLE XV

F_1	Weight of function-input F_1 to 2	F_2	Sum of weights of all other inputs to 2	Sum of weights of all inputs to 2	F
1	$a_{12} = T_2$	1	$\geq T_2$	$\geq 2T_2 > T_2$	1
1	$a_{12} = T_2$	0	$< T_2$	$\geq T_2$	1
0	0	1	$\geq T_2$	$\geq T_2$	1
0	0	0	$< T_2$	$< T_2$	0

From Table XV it is seen that

$$F = F_1 + F_2$$

Thus part (a) of Theorem 8.2.3 is proved.

(b) Let us connect F_1 as a function-input to the last-stage element of F_2 with a weight $a_{12} > \sum_{i=1}^{n} a_{i(2)}$, where $\sum_{i=1}^{n} a_{i(2)}$ is the sum of the weights of all the other inputs to the last-stage element realizing F_2, with the threshold value increased such that

$$T_2' = T_2 + a_{12}$$

where T_2' = new threshold value

T_2 = original threshold value

as shown in Fig. 9. The truth conditions of the functions are shown in Table XVI.

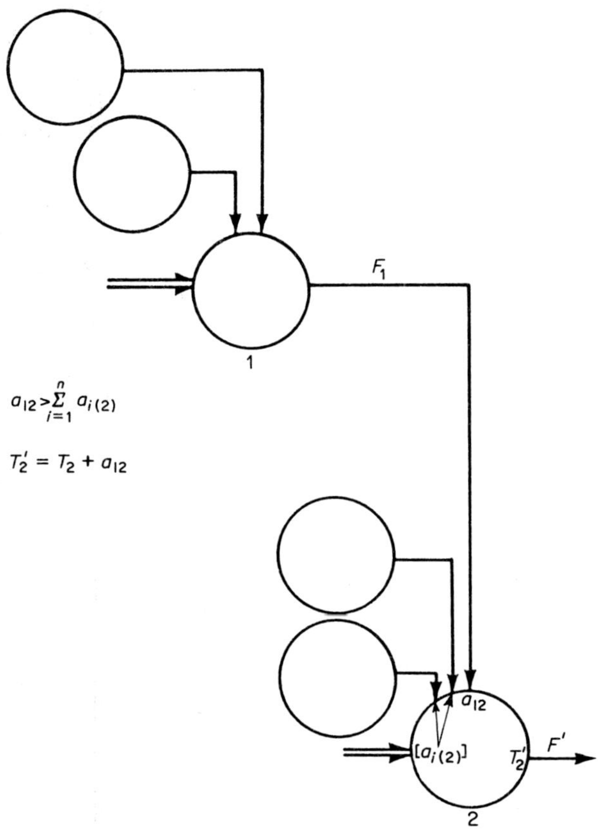

FIG. 9. Realization of $F = F_1 F_2$.

TABLE XVI

F_1	Weight of function-input F_1 to 2	F_2	Sum of weights of all other inputs to 2	Sum of weights of all inputs to 2	F
1	a_{12}	1	$\geq T_2$	$\geq T_2+a_{12} = T_2'$	1
1	a_{12}	0	$< T_2$	$< T_2+a_{12} = T_2'$	0
0	0	1	$\geq T_2$	$< a_{12} < T_2+a_{12} = T_2'$	0
0	0	0	$< T_2$	$< T_2 < T_2+a_{12} = T_2'$	0

From Table XVI it is seen that

$$F = F_1 F_2 = F$$

Thus part (b) of Theorem 8.2.3 is proved. Q.E.D.

Theorem 8.2.3 is important in that using it we not only can realize the combination of two arbitrary Boolean functions, but also can realize the combination of more than two functions at a time.

THEOREM 8.2.4. If an arbitrary Boolean function F of n variables can be expressed as any successive additive and/or multiplicative combination of m threshold functions, $F_{T1}, F_{T2}, \ldots, F_{Tm}$, then F can be realized with m threshold logic elements, or is m-realizable.

PROOF: This theorem is in fact a generalization of Theorems 8.2.1, 8.2.2, and 8.2.3. Any function, as a successive additive and/or multiplicative combination of threshold functions, by Theorems 8.2.1, 8.2.2, and 8.2.3, can be reduced, step by step, to combinations of fewer and fewer functions, and finally reduced to a single function F. The successive steps of reduction of the function leading to a single function are equivalent to the realization with threshold logic elements. Q.E.D.

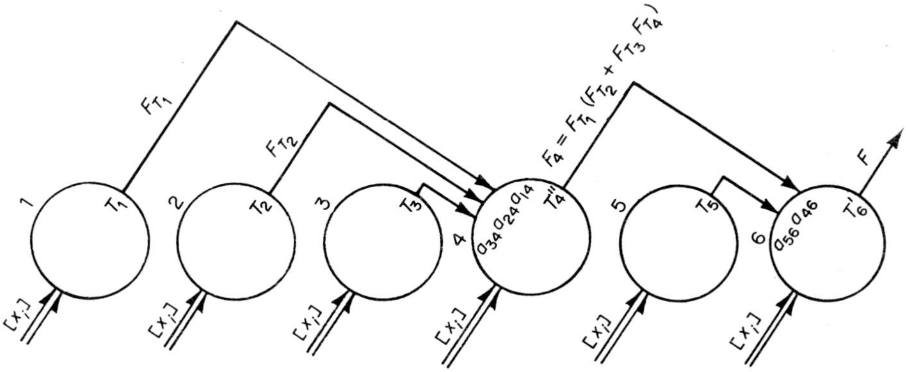

FIG. 10. Realization of F in Example 8.2–1.

Let us consider a general example.

EXAMPLE 8.2.1

$F = F_{T1}(F_{T2}+F_{T3}F_{T4})+F_{T5}F_{T6}$

The realization is shown in Fig. 10. The weights and threshold values may be chosen as follows:

$$a_{34} > \sum_{i=1}^{n} a_{i(4)}$$
$$T_4' = T_4+a_{34}$$
$$a_{24} = T_4' = T_4+a_{34}$$
$$a_{14} > a_{24}+a_{34}+\sum_{i=1}^{n} a_{i(4)}$$
$$T_4'' = T_4'+a_{14} = T_4+a_{14}+a_{34}$$
$$a_{56} > \sum_{i=1}^{n} a_{i(6)}$$
$$T_6' = T_6+a_{56}$$
$$a_{46} = T_6'$$

THEOREM 8.2.5. If an arbitrary Boolean function F of n variables can be expressed as a combination of a number of threshold functions, some or all of which have repeated occurrences, and let j be the number of threshold functions when every function with repeated occurrences is counted as one, and let k be the number of threshold functions when repeated occurrences of the same function are counted as different functions, then F can be realized with m threshold logic elements, such that

$$j \leq m \leq k$$

PROOF: That $m \leqslant k$ is obvious. Consider the threshold functions as k different functions. Then by Theorem 8.2.4, $m = k$.

Now for a threshold function F_T with repeated occurrences, it is sometimes possible to use only one threshold logic element to obtain various combinations with other threshold functions. When it is possible to use only one threshold logic element for each threshold function with repeated occurrences in the given function, we have $m = j$. When some threshold functions with repeated occurrences each need only one threshold logic element for realization, but some other threshold functions with repeated occurrences each need more than one threshold logic element for realization, then we have $j < m < k$.

Thus we have shown that in different situations m may be equal to k, equal to j, or smaller than k but greater than j. Therefore $j \leq m \leq k$.

Q.E.D.

Let us consider an example in which a threshold function with repeated occurrences needs only one threshold logic element for its realization.

EXAMPLE 8.2–2.

$F = (F_{T1} + F_{T2})F_{T3} + F_{T1}F_{T4}$

In this expression, F_{T1} has two occurrences, but it needs only one threshold logic element for realization. Therefore F can be realized with 4 threshold logic elements, as shown in Fig. 11.

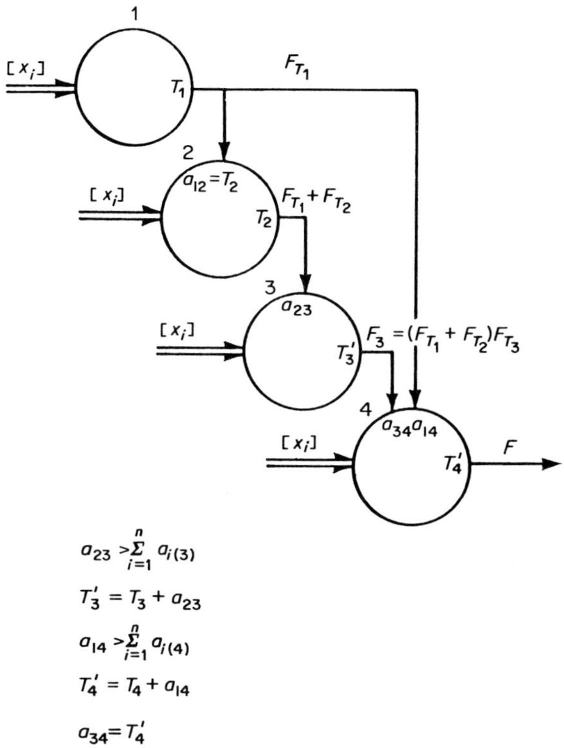

FIG. 11. Realization of F in Example 8.2–2.

A realization of a Boolean function with threshold logic elements is *minimal* iff the number of threshold logic elements used in the realization is a minimum.

It is to be noted that the previous five theorems do not give the minimum number of elements required, but rather an upper bound of minimum number of elements.

As minimal realization is difficult to obtain, we only attempt to find a good realization close to minimal. A relatively simple way is to express the function F as a sum of m threshold functions or as a product of m' threshold functions. To express F as a sum of threshold functions, we can first express F in the sum-of-product form and then group several terms together to form threshold functions. Similarly, to express F as a product of threshold functions we can first express F in the product-of-sum form and then group several factors together to form threshold functions.

It is of interest to note that when a function F is expressed as the sum of m threshold functions and when it is also expressed as the product of m' threshold functions, m is not necessarily equal to m'. m may be greater than, equal to, or smaller than m', depending on the function. Let us consider the following examples.

EXAMPLE 8.2–3

$$F = \bar{x}_1\bar{x}_2\bar{x}_3\bar{x}_4+\bar{x}_1\bar{x}_2\bar{x}_3x_4+\bar{x}_1\bar{x}_2\bar{x}_3\bar{x}_4+\bar{x}_1x_2x_3x_4+x_1\bar{x}_2x_3x_4 +x_1x_2x_3\bar{x}_4 \quad (8.2\text{–}1)$$

$$= (x_1+x_2+\bar{x}_3)(x_1+x_2+x_4)(x_1+\bar{x}_3+x_4)(x_2+\bar{x}_3+x_4) (\bar{x}_1+\bar{x}_2+x_3)(\bar{x}_1+\bar{x}_2+\bar{x}_4)(\bar{x}_1+x_3+\bar{x}_4)(\bar{x}_2+x_3+\bar{x}_4) \quad (8.2\text{–}2)$$

Equation (8.2–1) is the sum-of-product form of F and Equation (8.2–2) is the product-of-sum form. From Equation (8.2–1) it is easily seen that the sum of any two terms has some variables in both the uncomplemented and complemented forms. Thus the sum of any two terms is not a unate function, and therefore not a threshold function. So each term is a threshold function by itself, and there are altogether six threshold functions, or $m = 6$. From Equation (8.2–2) it is seen that the first four factors can be grouped to form a threshold function of four variables x_1, x_2, $\bar{x}_3$, and x_4, with $a_1 = a_2 = a_3 = a_4 = 1$ and $T = 2$. Thus

$$R[F(x_1, x_2, \bar{x}_3, x_4)] = (1, 1, 1, 1; 2)$$

The remaining four factors can also be grouped to form a threshold functions of four variables $\bar{x}_1$, $\bar{x}_2$, x_3, and $\bar{x}_4$, with $a_1 = a_2 = a_3 = a_4 = 1$ and $T = 2$. Or

$$R[F(\bar{x}_1, \bar{x}_2, x_3, \bar{x}_4)] = (1, 1, 1, 1; 2)$$

Therefore F can be realized with two threshold logic elements in the product form, or $F = F_{T1}F_{T2}$, and $m' = 2$.

EXAMPLE 8.2–4

$$F = x_1x_2\bar{x}_3+x_1x_2x_4+x_1\bar{x}_3x_4+x_2\bar{x}_3x_4+\bar{x}_1\bar{x}_2x_3+\bar{x}_1\bar{x}_2\bar{x}_4 +\bar{x}_1x_3\bar{x}_4+\bar{x}_2x_3\bar{x}_4$$

$$= (x_1+\bar{x}_2+\bar{x}_3+\bar{x}_4)(\bar{x}_1+x_2+\bar{x}_3+\bar{x}_4)(\bar{x}_1+\bar{x}_2+\bar{x}_3+x_4) (\bar{x}_1+x_2+x_3+x_4)(x_1+\bar{x}_2+x_3+x_4)(x_1+x_2+x_3+\bar{x}_4)$$

This function is the dual function of that in Example 8.2–3. In the sum-of-product form F is the sum of two threshold functions, or $m = 2$; but in the product-of-sum form F is the product of six threshold functions, or $m' = 6$. That m is not necessarily equal to m' will be utilized for compound synthesis, because we can always choose the form with the smaller number of threshold functions.

8.3 GENERAL STRUCTURE OF THRESHOLD-LOGIC NETWORKS

Now we shall study the same problem from the synthesis point of view.

If we consider only combinatorial networks in which there are no feedback connections, the most general network of m threshold logic elements is of the form shown in Fig. 12.[14] Threshold logic element 1,

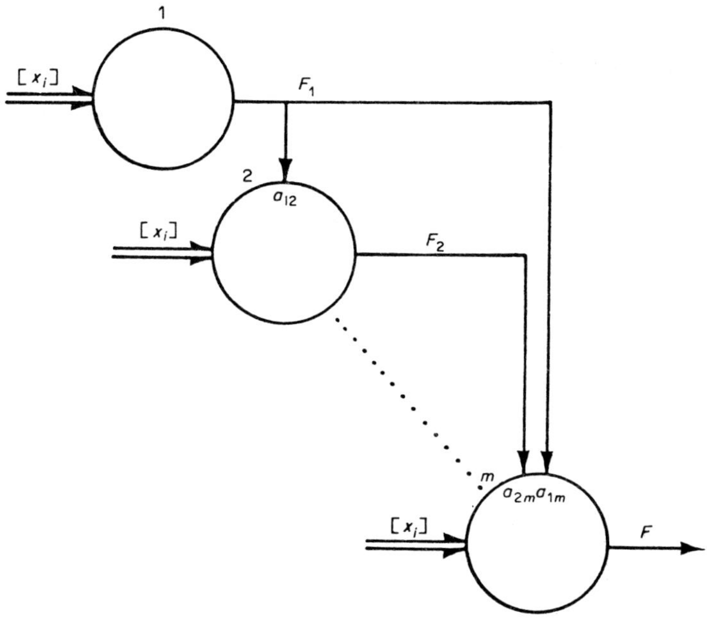

FIG. 12. General threshold-logic network.

called the 1*st* or 1*st-stage element*, has its output connected as a function-input to all the remaining $m-1$ threshold logic elements. Threshold logic element 2, the 2nd or 2nd-stage element, because of no feedback, cannot have its output connected as a function-input to element 1. But its output is connected as a function-input to the remaining $m-2$ elements. This reasoning extends to all the remaining elements, and the $(m-1)$th-stage element has its output connected as a function-input to the mth or m^{th}-stage (or last) element.

It is thus obvious that any other form of network is but a special case of this general network with some of the connections open, or with the weights of some function-inputs equal to zero.

We consider only single-output networks, with the output function of the last element being the required function. We shall find its expression in terms of its input functions. Let

$F_1, F_2, \ldots, F_m$ = output functions of threshold logic elements 1, 2,..., m, respectively. They are threshold functions of the inputs, but not necessarily of the variables $\{x_i\}$.

$F_{2/\bar{1}}$ = threshold function of element 2, when the function-input from the output of element 1 is 0.

$F_{2/1}$ = threshold function of element 2, when the function-input from the output of element 1 is 1.

$F_{3/\bar{1}\bar{2}}$ = threshold function of element 3, when the function-inputs from the outputs of both elements 1 and 2 are 0.

$F_{3/1\bar{2}}$ = threshold function of element 3, when the function-input from the output of element 1 is 1, but the function-input from the output of element 2 is 0.

$F_{3/\bar{1}2}$ = threshold function of element 3, when the function-input from the output of element 1 is 0, but the function-input from the output of element 2 is 1.

$F_{3/12}$ = threshold function of element 3, when the function-inputs from the outputs of both elements 1 and 2 are 1.

Now we can make some simplifications. If a_{ij}, the weight of F_i connected as a function-input to element j, is positive, then that $F_i = 1$ is equivalent to lowering the threshold value of element j by an amount a_{ij}, resulting in a threshold function containing more vertices. If a_{ij} is negative, then that $F_i = 1$ is equivalent to raising the threshold value of element j by an amount $|a_{ij}|$, resulting in a threshold function containing fewer vertices. The threshold functions thus obtained are obviously isobaric functions.

If we assume that all the weights are positive, we have

$$F_{j/i} \supset F_{j/\bar{i}}$$

This means that the output function obtained with a function-input equal to 1 always contains the output functions with the function-input equal to 0. This reasoning can be generalized to the kth-stage element with any number of function-inputs. For example, the 3rd-stage element has function-inputs from the 1st-stage and 2nd-stage elements. Then

$$F_{3/12} \supset \begin{matrix} F_{3/\bar{1}2} \\ F_{3/1\bar{2}} \end{matrix} \supset F_{3/\bar{1}\bar{2}}$$

$F_{3/\bar{1}2}$ and $F_{3/1\bar{2}}$ are also comparable. If $a_{13} > a_{23}$, then $F_{3/1\bar{2}} \supset F_{3/\bar{1}2}$; if $a_{23} > a_{13}$, then $F_{3/\bar{1}2} \supset F_{3/1\bar{2}}$.

Now we shall derive the output functions of networks of threshold logic elements for various values of m.

(1) $m = 1$

This case is trivial. The network reduces to a single element, and the output function is the threshold function of the element.

(2) $m = 2$

The network is shown in Fig. 13. The output function is

$$F_2 = F_1 F_{2/1} + \overline{F}_1 F_{2/\bar{1}} \tag{8.3–1}$$

Since for $a_{12} > 0$, $F_{2/1} \supset F_{2/\bar{1}}$,

$$F_2 = F_1 F_{2/1} + F_{2/\bar{1}} \tag{8.3–2}$$

From Equation (8.3–2) it is seen that according to Theorem 8.2.4, F_2 can be realized with 3 threshold logic elements, because there are three different threshold functions in the expression. However, F_2 actually can be realized with 2 threshold logic elements, as shown in Fig. 13,

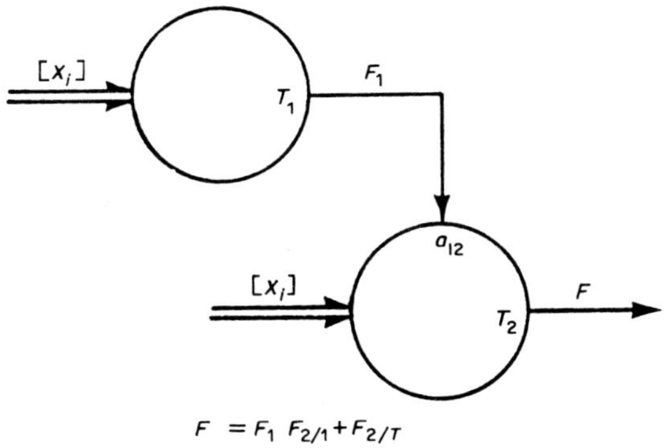

FIG. 13. Threshold-logic network of 2 elements.

because $F_{2/1}$ and $F_{2/\bar{1}}$ are isobaric functions. This gives a clue to compound synthesis. If we can express a general Boolean function as a combination of m threshold functions, such that some of them are isobaric functions, then we can realize the given function with k threshold logic elements, where $k \leq m$.

Equation (8.3–2) is in the sum-of-product form (in terms of the threshold functions). F can also be expressed in the product-of-sum form. The complement of F_2 is

$$\begin{aligned} \overline{F}_2 &= \overline{(F_1 F_{2/1} + F_{2/\bar{1}})} \\ &= (\overline{F}_1 + \overline{F}_{2/1})\overline{F}_{2/\bar{1}} \end{aligned} \tag{8.3–3}$$

Since

$F_{2/1} \supset F_{2/\bar{1}}$ and $\overline{F}_{2/\bar{1}} \supset \overline{F}_{2/1}$,

$$\overline{F}_2 = \overline{F}_1 \overline{F}_{2/\bar{1}} + \overline{F}_{2/1} \tag{8.3–4}$$

and

$$F_2 = (F_1 + F_{2/\bar{1}}) F_{2/1} \tag{8.3–5}$$

Equation (8.3–5) is in the product-of-sum form (in terms of threshold functions). It is useful because sometimes $m' < m$, and then it is preferable to start synthesis from the product-of-sum form.

(3) $m = 3$

The network is shown in Fig. 14.

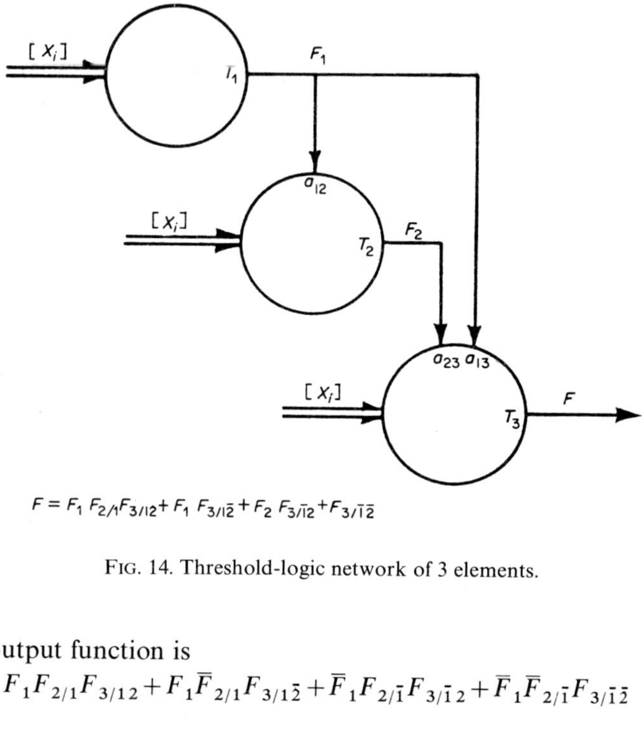

FIG. 14. Threshold-logic network of 3 elements.

The output function is

$$F_3 = F_1F_{2/1}F_{3/12} + F_1\bar{F}_{2/1}F_{3/1\bar{2}} + \bar{F}_1F_{2/\bar{1}}F_{3/\bar{1}2} + \bar{F}_1\bar{F}_{2/\bar{1}}F_{3/\bar{1}\bar{2}}$$

For

$a_{12} > 0, a_{13} > 0, a_{23} > 0,$

$$F_{2/1} \supset F_{2/\bar{1}},\ F_{3/12} \supset \begin{matrix} F_{3/1\bar{2}} \\ F_{3/\bar{1}2} \end{matrix} \supset F_{3/\bar{1}\bar{2}}$$

$$\bar{F}_{2/\bar{1}} \supset \bar{F}_{2/1},\ \bar{F}_{3/\bar{1}\bar{2}} \supset \begin{matrix} \bar{F}_{3/1\bar{2}} \\ \bar{F}_{3/\bar{1}2} \end{matrix} \supset F_{3/12}$$

Therefore

$$F_3 = F_1F_{2/1}F_{3/12} + F_1F_{3/1\bar{2}} + F_{2/\bar{1}}F_{3/\bar{1}2} + F_{3/\bar{1}\bar{2}} \tag{8.3–6}$$

Equation (8.3–6) is in the sum-of-product form. The complement of F_3 is

$$\begin{aligned} \bar{F}_3 &= (\bar{F}_1+\bar{F}_{2/1}+\bar{F}_{3/12})(\bar{F}_1+\bar{F}_{3/1\bar{2}})(\bar{F}_{2/\bar{1}}+\bar{F}_{3/\bar{1}2})\bar{F}_{3/\bar{1}\bar{2}} \\ &= \bar{F}_1\bar{F}_{2/\bar{1}}\bar{F}_{3/\bar{1}\bar{2}} + \bar{F}_1\bar{F}_{3/\bar{1}2} + \bar{F}_{2/1}\bar{F}_{3/1\bar{2}} + \bar{F}_{3/12} \end{aligned} \tag{8.3–7}$$

In the product-of-sum form, the output function is

$$F_3 = (F_1 + F_{2/\bar{1}} + F_{3/\bar{1}\bar{2}})(F_1 + F_{3/\bar{1}2})(F_{2/1} + F_{3/1\bar{2}})F_{3/12} \qquad (8.3\text{–}8)$$

The expressions for higher values of m can be easily derived in a similar manner and will not be given here. Equation (8.3–6) suggests a general method of compound synthesis from the sum-of-product form. The last term $F_{3/\bar{1}\bar{2}}$ is a threshold function and can be chosen from one of the threshold functions that add up to form F_3. The third term $F_{2/\bar{1}}F_{3/\bar{1}2}$ is the product of $F_{2/\bar{1}}$, a second threshold function, and $F_{3/\bar{1}2}$, an isobaric function of $F_{3/\bar{1}\bar{2}}$. The second term is the product of F_1, a third threshold function, and $F_{3/1\bar{2}}$, another isobaric function of $F_{3/\bar{1}\bar{2}}$. The first term is the product of F_1, the third threshold function, $F_{2/1}$, an isobaric function of $F_{2/\bar{1}}$, and $F_{3/12}$, another isobaric function of $F_{3/\bar{1}\bar{2}}$.

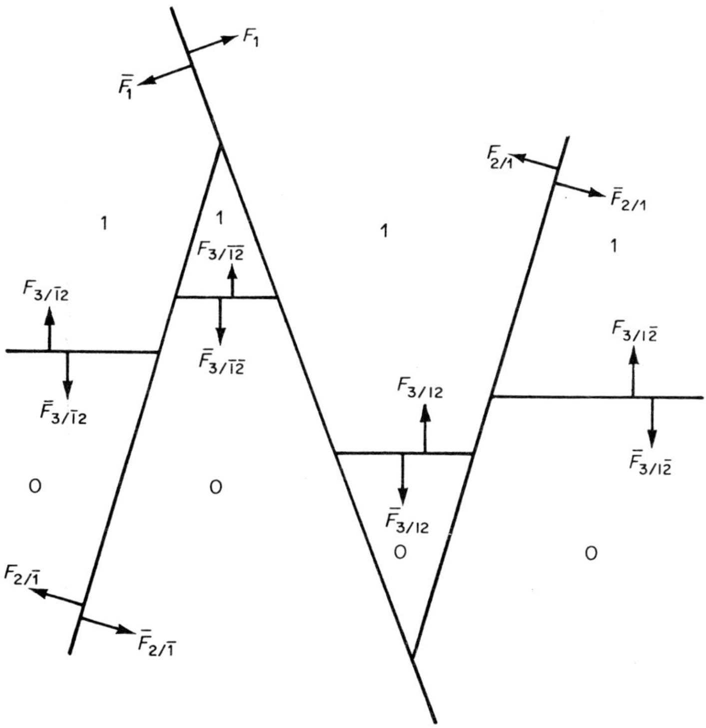

FIG. 15. Geometrical interpretation of 3-realization.

In terms of Winders' geometrical interpretation,[27] the division of the n-cube into regions by the threshold functions for a 3-realization ($m = 3$) is shown in Fig. 15.

The following example shows such a decomposition of a given function.

EXAMPLE 8.3–1

$$F = x_1x_2x_3+x_1x_2\bar{x}_4+x_2x_3\bar{x}_4+x_1\bar{x}_2\bar{x}_3x_4+\bar{x}_1\bar{x}_2x_3x_4$$
$$= (x_1x_2x_3+x_1x_2\bar{x}_4+x_2x_3\bar{x}_4)+(\bar{x}_1\bar{x}_2x_4+\bar{x}_2\bar{x}_3x_4)(x_1+x_3)$$
$$= x_2(x_1x_3+x_1\bar{x}_4+x_3\bar{x}_4)$$
$$+[\bar{x}_2x_4(\bar{x}_1+\bar{x}_3)][x_1+x_2+x_3+\bar{x}_4]$$
$$= F_{2/\bar{1}}+F_1F_{2/1}$$

where

$F_{2/\bar{1}}(x_1, x_2, x_3, x_4) = x_2(x_1x_3+x_1\bar{x}_4+x_3\bar{x}_4)$

$R[F_{2/\bar{1}}] = (1, 2, 1, 1; 4)$

$F_{2/1}(x_1, x_2, x_3, \bar{x}_4) = x_1+x_2+x_3+\bar{x}_4$

$R[F_{2/1}] = (1, 2, 1, 1; 1)$

$F_1(\bar{x}_1, \bar{x}_2, \bar{x}_3, x_4) = \bar{x}_2x_4(\bar{x}_1+\bar{x}_3)$

$R[F_1] = (1, 2, 1, 2; 5)$

$a_{12} = 4-1 = 3$

Thus F is realizable with two threshold logic elements. The realization is shown in Fig. 16.

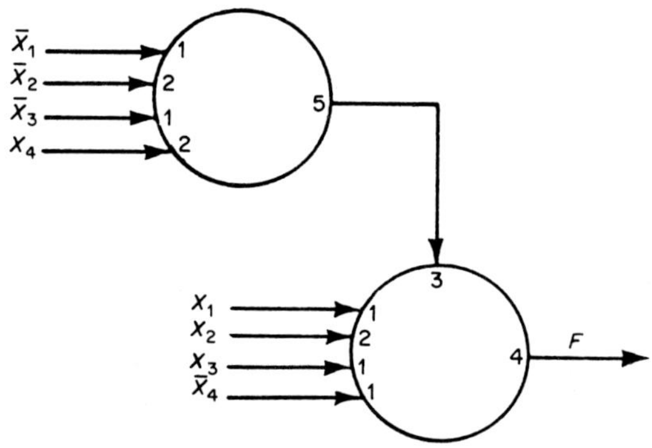

FIG. 16. Realization of F in Example 8.3–1.

8.4 SOME UNUSUAL SITUATIONS

From the above study of the structure of networks of threshold logic elements it is seen that it is an indispensable step to express a function or a part of a function as a sum or product of several threshold functions. Since unate functions can be tested by inspection, it is much easier to express a function as a sum or product of several unate functions. So we shall consider the decomposition of a function into unate functions first. Since the product-of-sum form is dual to sum-of-product form, we shall study the latter form only, and any theorem or reasoning for the latter can be easily extended to the former.

A minimum sum-of-product form or an irredundant normal disjunctive form usually contains a minimum number of terms with minimum number of literals. In general it can be expected that we can find from the minimum sum-of-product form or irredundant normal disjunctive form the function expressed as a sum of a minimum number of unate functions. However, this is not necessarily true, as shown by the following theorems.

THEOREM 8.4.1. A function in its minimum sum-of-product form is not necessarily the sum of a minimum number of unate functions.

PROOF: We shall prove this theorem by giving a counter example. Consider the following function F.

$$F = x_1\bar{x}_3 + \bar{x}_1 x_2 x_4 + \bar{x}_1\bar{x}_2 x_3 \qquad \text{(minimum sum-of-product form)}$$
$$= (x_1\bar{x}_3 + x_2\bar{x}_3 x_4) + (\bar{x}_1 x_3 x_4 + \bar{x}_1\bar{x}_2 x_3) \qquad \text{(non-minimum form)}$$

It is seen that in the minimum sum-of-product form F is the sum of three unate functions, but in the non-minimum form F is the sum of only two unate functions. Q.E.D.

THEOREM 8.4.2. A function in its minimum sum-of-product form is not necessarily the sum of a minimum number of threshold functions.

PROOF: We shall prove this theorem again by giving a counter example. In the example in the proof of Theorem 8.4.1, each unate function is also a threshold function. Q.E.D.

THEOREM 8.4.3. A function in its irredundant normal disjunctive form (indf) is not necessarily the sum of a minimum number of threshold functions.

PROOF: We shall prove this theorem again by giving a counter example. Consider the following function F.

$$F = x_1x_2x_3 + x_1x_2x_4 + x_1x_3x_5 + x_2\bar{x}_3\bar{x}_4x_5 \qquad \text{(indf)}$$
$$= (x_1x_2x_3 + x_1x_2x_4 + x_1x_2x_5 + x_1x_3x_5) + (x_2\bar{x}_3\bar{x}_4x_5) \qquad \text{(non-indf)}$$

It is seen that when in the indf, F is the sum of three threshold functions, but when not in the indf ($x_1x_2x_5$ is redundant) F is the sum of only two threshold functions. Q.E.D.

The unate function $(x_1x_2x_3 + x_1x_2x_4 + x_1x_3x_5)$ is not a threshold function, but is the sum of two threshold functions. The redundant term $x_1x_2x_5$ added contains four n-tuples: $x_1x_2x_3x_4x_5, x_1x_2x_3\bar{x}_4x_5, x_1x_2\bar{x}_3x_4x_5$, and $x_1x_2\bar{x}_3\bar{x}_4x_5$. The first three are contained in the unate function $(x_1x_2x_3 + x_1x_2x_4 + x_1x_3x_5)$. Now by adding the n-tuple $x_1x_2\bar{x}_3\bar{x}_4x_5$ to the unate function we obtain a threshold function. This shows that it is sometimes possible to obtain a threshold function from a unate function by adding redundant n-tuples not contained in this unate function, thus reducing the number of threshold functions required to form the given function.

THEOREM 8.4.4. An expression containing no non-prime implicants is not necessarily the sum of fewer number of threshold functions than an expression containing some non-prime implicants.

PROOF: We shall prove this theorem again by giving a counter example. Consider the following function F.

$F = \bar{x}_1\bar{x}_3+\bar{x}_2\bar{x}_4+x_1x_3+x_2x_4$ (minimum sum-of-product form containing no non-prime implicants)

$= (\bar{x}_1\bar{x}_3+\bar{x}_1\bar{x}_2\bar{x}_4+\bar{x}_2\bar{x}_3\bar{x}_4)+(x_1x_3+x_1x_2x_4+x_2x_3x_4)$ (non-minimum sum-of-product form containing some non-prime implicants).

It is seen that when in the minimum sum-of-product form containing no non-prime implicants, F is the sum of four threshold functions, but when in the non-minimum sum-of-product form containing some non-prime implicants, F is the sum of only two threshold functions.

Q.E.D.

Corresponding to the above four theorems there exist theorems for the product-of-sum form too.

8.5 DECOMPOSITION INTO UNATE FUNCTIONS

To express a function F as a sum of minimum number of unate functions, first we minimize F and express it in an indf. As pointed out in Section 8.4, if there are more than one indf's, the minimum form is not necessarily the best form for our purpose, and we have to try every prime implicant so as to choose the one that is the sum of a minimum number of unate functions.

To find unate functions, we can test systematically the pairwise unateness between every pair of terms of the function.

If two terms (factors) of a function have no variable appearing in both complemented and uncomplemented form, it is said that there is a *pairwise unateness* between these two terms.

THEOREM 8.5.1. A function is a unate function iff it can be expressed in a form such that there is pairwise unateness between every two terms (factors).

PROOF: This is obvious according to the definition of unate functions.

Q.E.D.

To find pairwise unateness between every two terms (factors) of a function systematically we can construct a table similar to that used for the check of compatibility between states of a sequential machine.

To illustrate this method let us consider an example.

EXAMPLE 8.5–1

$F(x_1, x_2, x_3, x_4, x_5) = \{1, 5, 6, 8, 9, 12, 13, 15, 16, 17, 20, 22, 23, 25, 26, 27\}$

F has three indf's

$$F = \bar{x}_1 x_2 \bar{x}_4 + \bar{x}_1 \bar{x}_4 \bar{x}_5 + x_1 x_3 x_4 + x_2 x_3 x_5 + \bar{x}_2 x_3 x_4 \bar{x}_5$$

$$+ \begin{cases} (x_1 \bar{x}_2 \bar{x}_3 \bar{x}_4 + x_1 \bar{x}_2 \bar{x}_4 \bar{x}_5) & \text{(form } A\text{)} \\ \text{or} & \\ (x_1 \bar{x}_2 \bar{x}_3 \bar{x}_4 + x_1 \bar{x}_2 x_3 \bar{x}_5) & \text{(form } B\text{)} \\ \text{or} & \\ (\bar{x}_2 \bar{x}_3 \bar{x}_4 x_5 + x_1 \bar{x}_2 \bar{x}_4 \bar{x}_5) & \text{(form } C\text{)} \end{cases}$$

In each indf there are seven terms, and the terms 6 and 7 have three alternatives. Now we check for pairwise unateness by forming the following three tables. The terms are denoted by 1, 2, ..., 7, respectively, according to the order of appearance in the above expressions. If between the i and the j terms there is a pairwise unateness, we put a $\surd$ mark in the cell at the intersection of the i row and the j column.

Indf A

	7	6	5	4	3	2
6	√					
5						
4						
3			√	√		
2				√		
1				√		√

Indf B

	7	6	5	4	3	2
6						
5	√					
4						
3	√		√	√		
2				√		
1				√		√

Indf C

	7	6	5	4	3	2
6						
5						
4						
3			√	√		
2				√		
1		√		√		√

Now we write down all the pairwise unateness pairs. If a term has no pairwise unateness with any other term, we keep the term itself as a unate function. Then we collect terms unate to one another to form unate functions with as many terms as possible. Finally we form a set or sets of unate functions such that the sum of the unate functions in each set is the given function. If there are more than one set, we choose the set with the minimum number of unate functions.

For Form A,
Pairwise-unateness pairs: (1, 2)(1, 4)(2,4)(3, 4)(3, 5)(6, 7)
Set of unate functions: {(1, 2, 4)(3, 5)(6, 7)}
For Form B,
Pairwise-unateness pairs: (1, 2)(1, 4)(2, 4)(3, 4)(3, 5)(3, 7)(5, 7)(6)

Set of unate functions: {(1, 2, 4)(3, 5, 7)(6)}
For Form C,
Pairwise-unateness pairs: (1, 2)(1, 4)(1, 6)(2, 4)(3, 4)(3, 5)(7)
Sets of unate functions: {(1, 2, 4)(3, 5)(6)(7)}
{(1, 6)(2, 4)(3, 5)(7)}

Indf A is the sum of three unate functions; indf B is the sum of three unate functions; indf C is the sum of four unate functions. So we discard C and may choose either A or B. In case of a tie like this, it is desirable to choose the one with the minimum number of threshold functions. In this example it happens that each unate function is a threshold function and for both forms A and B the number of threshold functions are also the same.

8.6 DECOMPOSITION INTO THRESHOLD FUNCTIONS

To decompose a function into the sum (product) of a minimum number of threshold functions, we first decompose the function into a sum (product) of minimum number of unate functions.

THEOREM 8.6.1. If m_U is the minimum number of unate functions into which a function F can be decomposed, then F can be decomposed into the sum of m_T threshold functions such that

$$m_T \geq m_U$$

PROOF: This is obvious because every threshold function is also a unate function, but not vice versa. Q.E.D.

There is a corresponding theorem for the product-of-sum form. By applying existing methods for testing of 1-realizability we test each unate function to see if it is also a threshold function. If a function F_U is also a threshold function, no more improvement can be made. If F_U is not a threshold function, by Theorems 8.4.3 and 8.4.4, it might become a threshold function with a few vertices added to it or taken away from it. So we must try to add some redundant terms to F_U to see if F_U can be changed into a threshold function, as shown in the example in the proof of Theorem 8.4.3. We must also try to break some term or terms of F_U into non-prime implicants and take away some of the redundant non-prime implicants to see if F_U can be changed into a threshold function, as shown in the example in the proof of Theorem 8.4.4. If by so doing F_U still cannot be changed into a threshold function, then we have to decompose F_U into a sum of several threshold functions.

When the number of variables and/or the number of terms in a unate function is not large, it is comparatively easy to test for 1-realizability and to decompose it into a sum of threshold functions.

One systematic method of decomposition is to decompose a function along one or more variables, as follows:

$$F = x_1F_1 + F_2$$
$$F = x_1x_2F_1 + x_1F_2 + x_2F_3 + F_4$$

This method does not necessarily result in the minimum number of threshold functions, although it is systematic. By successive decomposition we can reduce the number of variables in each of the resulting functions to as few as is desirable. The farthest we may go is where the resulting functions have three variables, because a unate function of three or fewer variables is always a threshold function.

Since a unate function usually does not contain many terms (factors), we often can decompose it into the sum (product) of several threshold functions by inspection.

8.7 SYNTHESIS PROCEDURE

From Section 8.3 it is seen that if we can express an arbitrary Boolean function in the form of Equation (8.3–6) or (8.3–8), then the function can be realized with m threshold logic elements, connected in the form of the general structure as shown in Fig. 12.

If trial and error is permitted and if we can have the best choice, then we can formulate a synthesis procedure as follows:

(1) Express the given function F as the sum of a minimum number, m, of threshold functions. Also express F as the product of a minimum number, m', of threshold functions. We are assured that F can be realized with min (m, m') threshold logic elements.

(2) If $m > m'$, we start realization from the product-of-sum form. If $m < m'$, we start realization from the sum-of-product form. (Actually these two procedures are dual to each other and are quite similar. So we just describe one procedure and omit the other.) Suppose $m < m'$, and we start from the sum-of-product form.

Choose a threshold function containing not too many vertices to be $F_{k/\bar{1}\ldots\overline{(k-1)}}$, the function of the kth or last-stage element. The reason for choosing a function containing not too many vertices is that we shall try to find its isobaric functions that will contain $F_{k/\bar{1}\ldots\overline{(k-1)}}$.

(3) Try to find two or more of the remaining $(k-1)$ threshold functions, such that their sum, expressed in product-of-sum form, is the product of an isobaric function of $F_{k/\bar{1}\ldots\overline{(k-1)}}$ and a new threshold function, $F_{(k-1)/\bar{1}\ldots\overline{(k-2)}}$. If such a sum cannot be found then choose another threshold function as $F_{(k-1)/\bar{1}\ldots\overline{(k-2)}}$.

(4) Try to find two or more of the remaining threshold functions, such that their sum, expressed in product-of-sum form, is the product of an

isobaric function of $F_{k/\bar{1}\ldots\overline{(k-1)}}$ and another new threshold function, $F_{(k-2)/\bar{1}\ldots\overline{(k-3)}}$. Then try to find two or more of the remaining threshold functions, such that their sum, expressed in product-of-sum form, is the product of an isobaric function of $F_{k/\bar{1}\ldots\overline{(k-1)}}$, an isobaric function of $F_{(k-1)/\bar{1}\ldots\overline{(k-2)}}$, and $F_{(k-2)/\bar{1}\ldots\overline{(k-3)}}$.

In any of these steps, if no such a sum can be found, proceed to the next step.

(5) Continue this process to find new threshold functions and products of threshold functions until all the threshold functions in the original sum-of-product form are covered.

In this manner we can find a realization with at most m threshold logic elements. By trying a few different choices we may be quite close to the minimal realization.

The following example will not show the details of synthesis procedure, but only the results of decomposition of a certain function into threshold functions.

EXAMPLE 8.7–1

$$F(x_1, x_2, x_3, x_4, x_5) = \{0, 2, 7, 11, 13, 14, 17, 18, 19, 20, 21, 22, 24, 26\}$$

$$\begin{aligned}
F &= \bar{x}_1\bar{x}_2\bar{x}_3\bar{x}_5 + \bar{x}_1x_2x_3\bar{x}_4x_5 + \bar{x}_1x_2x_3x_4x_5 + \bar{x}_1\bar{x}_2x_3x_4x_5 + \bar{x}_1x_2\bar{x}_3x_4x_5 \\
&\quad + x_1\bar{x}_2x_4\bar{x}_5 + x_1\bar{x}_2\bar{x}_3x_5 + x_1\bar{x}_2x_3\bar{x}_4 + x_1x_2\bar{x}_3\bar{x}_5 \qquad \text{(indf)} \\
&= \bar{x}_1\bar{x}_2\bar{x}_3\bar{x}_5 + [(\bar{x}_1+\bar{x}_2+\bar{x}_5)(\bar{x}_1+\bar{x}_2+\bar{x}_3)(\bar{x}_1+\bar{x}_3+\bar{x}_4+\bar{x}_5) \\
&\qquad (\bar{x}_2+\bar{x}_3+\bar{x}_4+\bar{x}_5)] \\
&\quad [(x_1+x_2+x_3)(x_1+x_2+x_4)(x_1+x_2+x_5)(x_1+x_3+x_4) \\
&\qquad (x_1+x_3+x_5)(x_1+x_4+x_5)(x_2+x_3+x_4+x_5)] \\
&= \bar{x}_1\bar{x}_2\bar{x}_3\bar{x}_5 + [\bar{x}_1\bar{x}_2 + \bar{x}_1\bar{x}_3 + \bar{x}_1\bar{x}_4 + \bar{x}_1\bar{x}_5 + \bar{x}_2\bar{x}_3 + \bar{x}_2\bar{x}_4 + \bar{x}_2\bar{x}_5 \\
&\qquad + \bar{x}_3\bar{x}_5]\cdot \\
&\quad [x_1x_2 + x_1x_3 + x_1x_4 + x_1x_5 + x_2x_3x_4 + x_2x_3x_5 + x_2x_4x_5 + x_3x_4x_5] \\
&= F_{2/\bar{1}} + F_{2/1}F_1
\end{aligned}$$

$$R[F_{2/\bar{1}}(\bar{x}_1, \bar{x}_2, \bar{x}_3, \bar{x}_4, \bar{x}_5)] = (3, 3, 2, 1, 2; 10)$$

$$R[F_{2/1}(\bar{x}_1, \bar{x}_2, \bar{x}_3, \bar{x}_4, \bar{x}_5)] = (3, 3, 2, 1, 2; 4)$$

$$R[F_1(x_1, x_2, x_3, x_4, x_5)] = (2, 1, 1, 1, 1; 3)$$

$$a_{12} = 10 - 4 = 6$$

The realization of F is shown in Fig. 17.

This example is very special and interesting in that in the indf F is the sum of nine threshold functions, but after taking away the term $\bar{x}_1\bar{x}_2\bar{x}_3\bar{x}_5$ the remaining terms become the product of two threshold functions, one of which is isobaric to $\bar{x}_1\bar{x}_2\bar{x}_3\bar{x}_5$.

The success of the above synthesis procedure depends on the right choice of threshold functions. Besides, the searching for two or more

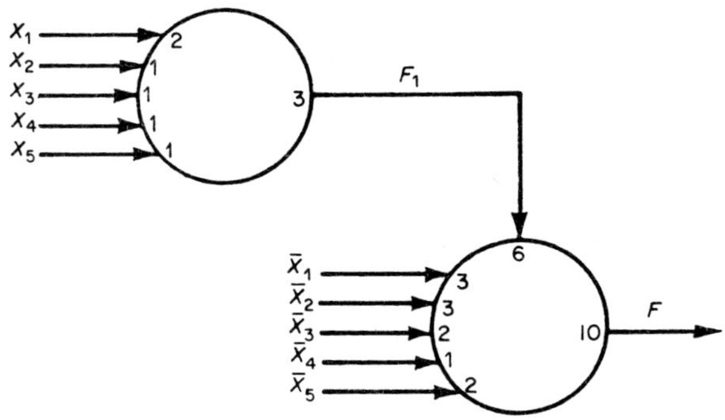

FIG. 17. Realization of F in Example 8.7–1.

threshold functions whose sum is the product of a new threshold function and isobaric functions of threshold functions already chosen needs a lot of cut and trial. The procedure can be simplified at the expense of number of threshold functions. In the following is given a more practical procedure for synthesis.

Step 1. Express the given function F as a sum of a minimum number, m, of threshold functions. Also express F as a product of a minimum number, m', of threshold functions. If $m < m'$, proceed to Step 2. If $m > m'$, proceed to Step 3. If $\min(m, m') < 2$, choose all the threshold functions and stop.

Step 2. Choose one of the m threshold functions in the sum-of-product form, preferably containing not too many vertices, as function F_k. Express the sum of the remaining $m-1$ threshold functions in the product-of-sum form. Take away any threshold function or functions that are isobaric to any threshold function or functions that have already been chosen from a sum-of-product form. Consider the remaining part of the function as a new given function and go back to Step 1.

Step 3. Choose one of the m' threshold functions in the product-of-sum form, preferably containing a large number of vertices, as function F_k. Express the product of the remaining $m'-1$ threshold functions in the sum-of-product form. Take away any threshold function or functions that are isobaric to any function or functions that have already been chosen from the product-of-sum forms. Consider the remaining terms as a new given function and go back to Step 1.

This practical procedure avoids the trouble of trying to find two or more terms such that their sum is the product of a new threshold function and isobaric functions of threshold functions already chosen. Emphasis is laid on the comparison of m and m'. Each time a threshold function is chosen, the remaining function is expressed in both the sum-of-product and the product-of-sum forms so that a comparison of m and m' may be made. At the same time any function isobaric to threshold functions already chosen are taken away, with a view to reducing the number of threshold functions which compose the remaining part of the function.

The following example will illustrate this procedure.

EXAMPLE 8.7–2

$F(x_1, x_2, x_3, x_4, x_5) = \{1, 4, 7, 10, 11, 14, 15, 18, 19, 20, 23, 24, 26, 27, 29\}$

A sum-of-product form, which is incidentally also minimum, is

$$F = \bar{x}_1x_2x_4+\bar{x}_2x_3\bar{x}_4\bar{x}_5+\bar{x}_1\bar{x}_2\bar{x}_3\bar{x}_4x_5+x_1\bar{x}_3x_4+x_1x_2\bar{x}_3\bar{x}_5 +x_1x_2x_3\bar{x}_4x_5+\bar{x}_2x_3x_4x_5$$

Terms 1 to 6 are essential prime implicants. Term 7 is not essential, and may be replaced by two terms $\bar{x}_1x_3x_4x_5$ and $x_1\bar{x}_2x_4x_5$. However, these two terms do not result in a smaller value of m. Therefore we choose the minimum form.

Check for Pairwise Unateness

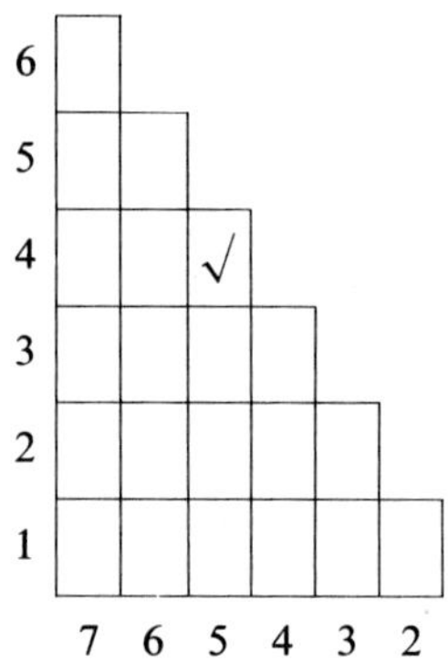

Pairwise unateness: (4, 5)

Set of unate functions: $\{(1)(2)(3)(4, 5)(6)(7)\}$

It happens that each unate function is also a threshold function, and the set of threshold functions is the same as the set of unate functions.

Therefore $m = 6$

One of the product-of-sum form of F is

$$F = (x_1+\bar{x}_2+x_4)(\bar{x}_1+\bar{x}_2+\bar{x}_3+x_5)(x_2+x_3+\bar{x}_4+\bar{x}_5)(\bar{x}_1+\bar{x}_2+\bar{x}_3+\bar{x}_4)(x_1+x_2+x_3+\bar{x}_4)(x_2+x_3+x_4+x_5)(x_2+\bar{x}_3+x_4+\bar{x}_5)$$

Check for Pairwise Unateness.

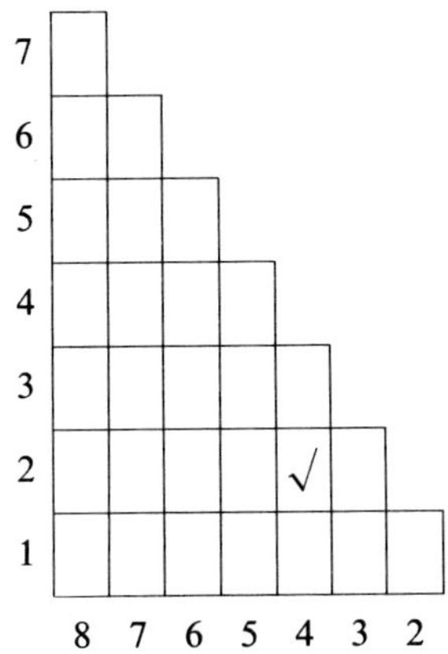

Pairwise unateness: (2, 4)

Set of unate functions: {(1)(2, 4)(3)(5)(6)(7)(8)}

Each unate function is also a threshold function, and the set of threshold functions is the same as the set of unate functions.

Therefore $m' = 7$

Since $m < m'$, we start synthesis from the sum-of-product form. As we do not know beforehand how many threshold functions F is composed of, we use subscripts $A, B, \ldots,$ for $k, k-1, \ldots$.

Choose $F_A = \bar{x}_1\bar{x}_2\bar{x}_3\bar{x}_4x_5$

Thus $F = F_A + F_{R(A)}$
where $F_{R(A)}$ is the remaining function after term F_A is taken away.

In product-of-sum form,

$$F_{R(A)} = (x_1+x_2+x_3)(x_1+\bar{x}_2+x_4)(x_2+x_4+\bar{x}_5)(x_2+x_3+x_4)$$
$$(x_3+x_4+\bar{x}_5)(x_2+\bar{x}_3+\bar{x}_4+x_5)(\bar{x}_1+\bar{x}_2+\bar{x}_3+\bar{x}_4)$$
$$(\bar{x}_1+\bar{x}_2+\bar{x}_3+x_5)$$

Now $F_{A'} = (\bar{x}_1+\bar{x}_2+\bar{x}_3+\bar{x}_4)(\bar{x}_1+\bar{x}_2+\bar{x}_3+\bar{x}_5)$ is isobaric to F_A. So take away this factor from $F_{R(A)}$, and the remaining function is

$$F_{R(A')} = (x_1+x_2+x_3)(x_1+\bar{x}_2+x_4)(x_2+x_4+\bar{x}_5)$$
$$(x_1+x_3+x_4)(x_3+x_4+\bar{x}_5)(x_2+\bar{x}_3+\bar{x}_4+x_5)$$

Check for Pairwise Unateness.

5					
4		√			
3		√	√		
2		√			
1		√	√	√	
	6	5	4	3	2

Pairwise unateness: (1, 3)(1, 4)(1, 5)(2, 5)(3, 4)(3, 5)(4, 5)
Set of unate functions: {(1, 3, 4, 5)(2)(6)}

$F_{R(A')}$ is the product of three unate functions. $(x_1+x_2+x_3)(x_2+x_4+\bar{x}_5)(x_2+x_3+x_4)(x_3+x_4+\bar{x}_5)$ is not a threshold function, but is the product of two threshold functions. Therefore $F_{R(A')}$ is the product of four threshold functions, or $m' = 4$.

In the sum-of-product form,

$F_{R(A')} = x_2x_4+x_3x_4x_5+x_1\bar{x}_3x_4+x_1x_2x_3+x_1x_2\bar{x}_5+\bar{x}_2x_3\bar{x}_4\bar{x}_5$

Check for Pairwise Unateness.

5					
4		√			
3		√			
2			√		
1		√	√	√	√
	6	5	4	3	2

Pairwise unateness: (1, 2)(1, 3)(1, 4)(1, 5)(2, 4)(3, 5)(4, 5)
Sets of unate functions: {(1, 2, 4)(3, 5)(6)}
{(1, 3, 5)(2, 4)(6)}
{(1, 4, 5)(2)(3)(6)}

The first set consists of three unate functions, but of five threshold functions. The second set consists of three unate functions, but of five threshold functions. The third set consists of four unate functions and also of four threshold functions. So $m = 4$.

Since $m = m'$, we can work on either form. Now we work on the product-of-sum form.

Choose $F_B = x_2+\bar{x}_3+x_4+x_5$

Then $F = F_A+F_{A'}F_BF_{R(B)}$

where $F_{R(B)}$ is the remaining function after F_B is taken away.

Now for $F_{R(B)}$, $m' = 3$

In sum-of-product form,

$F_{R(B)} = x_1x_4+x_2x_4+x_3x_4+x_1x_2x_3+x_1x_2\bar{x}_5+\bar{x}_2x_3\bar{x}_5$

Check for Pairwise Unateness.

5					
4		√			
3	√	√	√		
2		√	√	√	
1	√	√	√	√	√
	6	5	4	3	2

Pairwise unateness: (1, 2)(1, 3)(1, 4)(1, 5)(1, 6)(2, 3)
(2, 4)(2, 5)(3, 4)(3, 5)(3, 6)(4, 5)

Sets of unate functions: {(1, 2, 3, 4, 5)(6)}
{(1, 2, 4, 5)(3, 6)}
{(2, 3, 4, 5)(1, 6)}
{(2, 4, 5)(1, 3, 6)}

It is found that the sum of terms 1, 2, 3, 4, and 5 is a threshold function. So the first set consists of two unate functions and also of two threshold functions, or $m = 2$.

As soon as we obtain a sum or product of two threshold functions, the realization is practically completed, because no further simplification can be made.

Choose $F_C = x_1x_4+x_2x_4+x_3x_4+x_1x_2x_3+x_1x_2\bar{x}_5$

$F_D = \bar{x}_2x_3\bar{x}_5$

Now F can be expressed as follows:

$$F = F_A+F_{A'}F_B(F_C+F_D)$$

$$= \bar{x}_1\bar{x}_2\bar{x}_3\bar{x}_4x_5+(\bar{x}_1+\bar{x}_2+\bar{x}_3+\bar{x}_4x_5).(x_2+\bar{x}_3+\bar{x}_4+x_5).$$

$$[(x_1x_4+x_2x_4+x_3x_4+x_1x_2x_3+x_1x_2\bar{x}_5)+\bar{x}_2x_3\bar{x}_5]$$

$R[F_A(\bar{x}_1, \bar{x}_2, \bar{x}_3, \bar{x}_4, x_5)] = (2, 2, 2, 1, 1; 8)$

$R[F_{A'}(\bar{x}_1, \bar{x}_2, \bar{x}_3, \bar{x}_4, x_5)] = (2, 2, 2, 1, 1; 2)$

$$R[F_B(x_2, \bar{x}_3, \bar{x}_4, x_5)] = (1,1,1,1;1)$$
$$R[F_C(x_1, x_2, x_3, x_4, \bar{x}_5)] = (3,3,2,5,1;7)$$
$$R[F_D(\bar{x}_2, x_3, \bar{x}_5)] = (1,1,1;3)$$
$$a_{DC} = T_C = 7$$
$$A_{CB} = 5 > \sum a_{i(B)}$$
$$T'_B = T_B + a_{CB} = 1+5 = 6$$
$$a_{BA} = T_A - T_{A'} = 8-2 = 6$$

The realization of F is shown in Fig. 18.

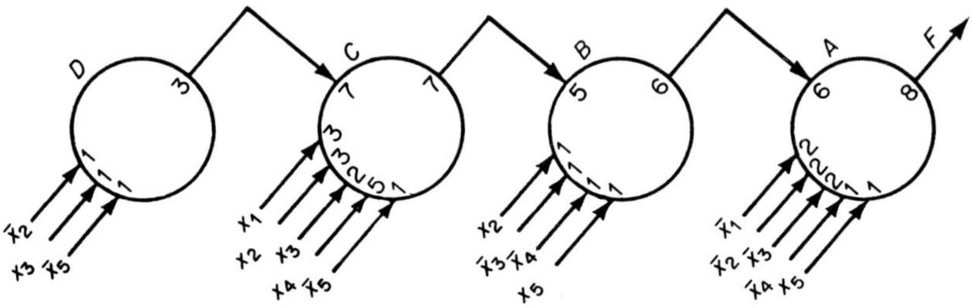

FIG. 18. Realization of F in Example 8.7–2.

8.8 SYNTHESIS OF THREE-INPUT MAJORITY GATE NETWORKS

A *three-input majority gate* is one of the simplest types of threshold logic elements, and enjoys its popular use as a vote-taker in triplicate circuits for reliable computation, and also as a universal building block in threshold-logic networks. As mentioned in Section 8.1, a large amount of study has been made for the realization of an arbitrary switching function with a network composed exclusively of three-input majority gates. Here we shall describe briefly only the method of Miyata,[20] which seems to be simple, straightforward, and systematic.

The output of a three-input majority gate having inputs x_1, x_2, and x_3 may be written as $x_1 \# x_2 \# x_3$, where

$$x_1 \# x_2 \# x_3 = x_1x_2 + x_2x_3 + x_3x_1 = x_1(x_2+x_3)+x_2x_3$$

If $x_2x_3 = 0$, then

$$x_1 \# x_2 \# x_3 = x_1(x_2+x_3) \qquad (8.8\text{–}1)$$

If $x_2+x_3 = 1$, then

$$x_1 \# x_2 \# x_3 = x_1 + x_2x_3 \qquad (8.8\text{–}2)$$

Now if a given switching function F has the following form

$$F = AB \qquad (8.8\text{–}3)$$

and one factor of F, for instance B, can be decomposed as

$$\begin{aligned} B &= C+D \\ CD &= 0 \end{aligned} \qquad (8.8\text{–}4)$$

then we have

$$F = AB = A(C+D)+CD = A \# C \# D \qquad (8.8\text{–}5)$$

Similarly, if F has the form

$$F = G+H \qquad (8.8\text{–}6)$$

and one term of F, for instance H, can be decomposed as

$$\begin{aligned} H &= KL \\ K+L &= 1 \end{aligned} \qquad (8.8\text{–}7)$$

then we have

$$F = G+H = G(K+L)+KL = G \# K \# L \qquad (8.8\text{–}8)$$

To realize F in the form of Equation (8.8–3), Equation (8.8–4) should be satisfied. It will be sufficient to decompose B to a sum of two terms along one of the variables, say x_1. Thus we have

$$B = x_1 B_{x_1} + \bar{x}_1 B_{\bar{x}_1} \qquad (8.8\text{–}9)$$

Similarly, to realize F in the form of Equation (8.8–6), Equation (8.8–7) should be satisfied. It will be sufficient to decompose H into a product of two factors. Thus we have

$$H = (x_1 + H_{\bar{x}_1})(\bar{x}_1 + H_{x_1}) \qquad (8.8\text{–}10)$$

Now we can realize a given function F in successive levels of product form systematically as follows. First we consider F as $1.F$ and decompose the second factor F to a sum of two terms along one of the variables, say x_1, as

$$\begin{aligned} F &= x_1 F_{x_1} + \bar{x}_1 F_{\bar{x}_1} & (8.8\text{–}11) \\ &= 1.x_1 F_{x_1} + 1.\bar{x}_1 F_{\bar{x}_1} + x_1 F_{x_1}.\bar{x}_1 F_{\bar{x}_1} & \\ &= 1 \# x_1 F_{x_1} \# \bar{x}_1 F_{\bar{x}_1} & (8.8\text{–}12) \end{aligned}$$

Since $x_1 F_{x_1}$ and $\bar{x}_1 F_{\bar{x}_1}$ are in the product form of Equation (8.8–3), F_{x_1} and $F_{\bar{x}_1}$ each can be decomposed along another variable, say x_2. Thus we have

$$F_{x_1} = x_2 F_{x_1 x_2} + \bar{x}_2 F_{x_1 \bar{x}_2} \qquad (8.8\text{–}13)$$

$$x_1 F_{x_1} = x_1 \# x_2 F_{x_1 x_2} \# \bar{x}_2 F_{x_1 \bar{x}_2} \qquad (8.8\text{–}14)$$

$$F_{\bar{x}_1} = x_2 F_{\bar{x}_1 x_2} + \bar{x}_2 F_{\bar{x}_1 \bar{x}_2} \tag{8.8–15}$$

$$\bar{x}_1 F_{\bar{x}_1} = \bar{x}_1 \ \# \ x_2 F_{\bar{x}_1 x_2} \ \# \ \bar{x}_2 F_{\bar{x}_1 \bar{x}_2} \tag{8.8–16}$$

$F_{x_1 x_2}$, $F_{x_1 \bar{x}_2}$, $F_{\bar{x}_1 x_2}$, and $F_{\bar{x}_1 \bar{x}_2}$ can be further decomposed along a third variable, say x_3, and the process of decomposition or expansion can be continued, until all the reduced functions become either a 0, a 1, or a single variable. Then the given function F will be realized by a network of three-input majority gates in the product form. The general structure of such a realization is shown in Fig. 19.

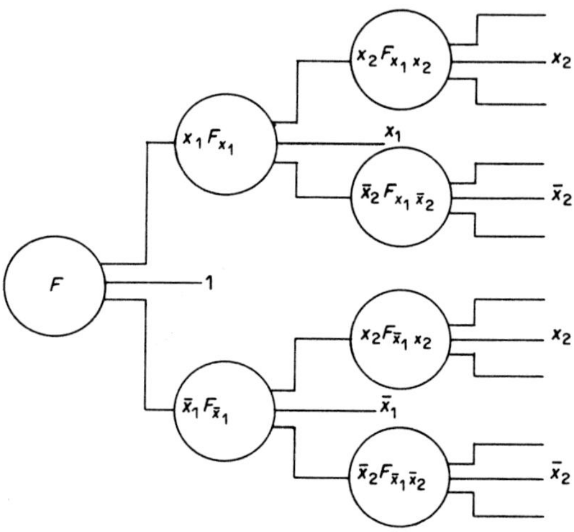

FIG. 19. Three-input majority gate network in product form.

We can also realize a given function F in successive levels of sum form systematically as follows. First we consider F as $0+F$, and decompose F to a product of two terms along one of the variables, say x_1, as

$$\begin{aligned} F &= (x_1 + F_{\bar{x}_1})(\bar{x}_1 + F_{x_1}) && (8.8–17) \\ &= 0(x_1 + F_{\bar{x}_1}) + 0(\bar{x}_1 + F_{x_1}) + (x_1 + F_{\bar{x}_1})(\bar{x}_1 + F_{x_1}) \\ &= 0 \ \# \ (x_1 + F_{\bar{x}_1}) \ \# \ (\bar{x}_1 + F_{x_1}) && (8.8–18) \end{aligned}$$

Since $(x_1 + F_{\bar{x}_1})$ and $(\bar{x}_1 + F_{x_1})$ are in the sum form of Equation (8.8–6), $F_{\bar{x}_1}$ and F_{x_1} each can be decomposed along another variable, say x_2. Thus we have

$$F_{\bar{x}_1} = (x_2 + F_{\bar{x}_1 \bar{x}_2})(\bar{x}_2 + F_{\bar{x}_1 x_2}) \tag{8.8–19}$$

$$x_1 + F_{\bar{x}_1} = x_1 \ \# \ (x_2 + F_{\bar{x}_1 \bar{x}_2}) \ \# (\bar{x}_2 + F_{\bar{x}_1 x_2}) \tag{8.8–20}$$

$$F_{x_1} = (x_2 + F_{x_1 \bar{x}_2})(\bar{x}_2 + F_{x_1 x_2}) \tag{8.8–21}$$

$$\bar{x}_1 + F_{x_1} = x_1 \ \# \ (x_2 + F_{x_1 \bar{x}_2}) \ \# \ (\bar{x}_2 + F_{x_1 x_2}) \tag{8.8–22}$$

The decomposition or expansion stops when all the reduced functions are either a 0, a 1, or a single variable. Then F will be realized by a network of three-input majority gates in the sum form. The general structure of such a realization is shown in Fig. 20.

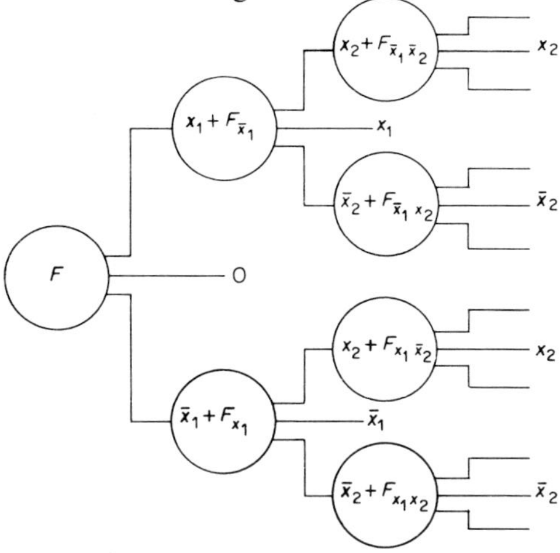

FIG. 20. Three-input majority gate network in sum form.

It is to be noted that in the decomposition or expansion, the order of separation of the variables, x_1, x_2, ..., may be selected arbitrarily, although the decompositions may result in different number of elements required.

One disadvantage of this method is that it does not recognize the function $AB+BC+CA$ directly, and therefore results in a larger number of elements than minimally required. However, whenever a function of the form $x_1x_2+x_2x_3+x_3x_1$ occurs, it is easily recognized by inspection, and can be represented as $x_1 \# x_2 \# x_3$ without further decomposition.

The following example illustrates the realization method essentially in the product form.

EXAMPLE 8.8–1

$$F = x_1x_3x_4+x_1\bar{x}_2x_3+x_1\bar{x}_2x_4+\bar{x}_1x_2x_3\bar{x}_4+\bar{x}_1x_2\bar{x}_3x_4$$

We select the order of separation of the variables as x_1, x_2, x_3', and x_4. Thus we have

$$\begin{aligned}
F_{x_1} &= x_3x_4+\bar{x}_2x_3+\bar{x}_2x_4 = \bar{x}_2 \# x_3 \# x_4 \\
F_{\bar{x}_1} &= x_2x_3\bar{x}_4+x_2\bar{x}_3x_4 \\
F_{\bar{x}_1x_2} &= x_3\bar{x}_4+\bar{x}_3x_4 \\
F_{\bar{x}_1\bar{x}_2} &= 0 \\
F_{\bar{x}_1x_2x_3} &= \bar{x}_4 \\
F_{\bar{x}_1x_2\bar{x}_3} &= x_4
\end{aligned}$$

The decomposition of F is shown in Fig. 21.

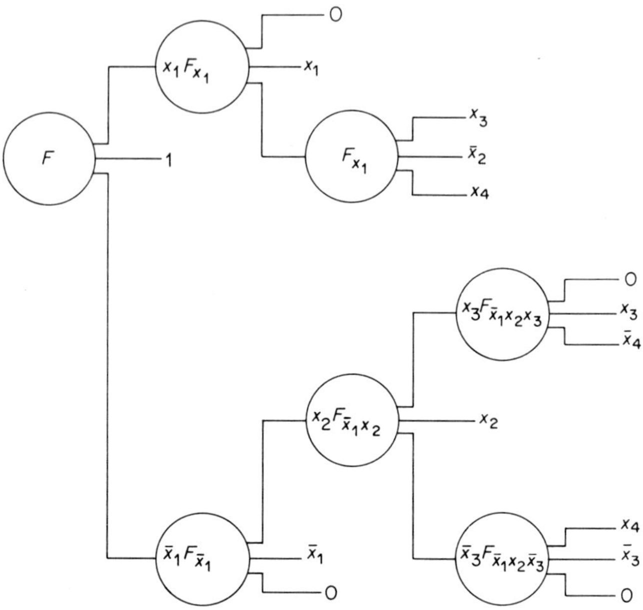

FIG. 21. Decomposition of F of Example 8.8–1.

The above method may be modified and applied in a more general case as follows:

Consider the case for which

$$F = AB = A(C+D) \tag{8.8–23}$$

but

$CD \neq 0$

Since $C+D = C+\bar{C}D$, we have

$$F = A(C+\bar{C}D) = A \# \bar{C} \# CD \tag{8.8–24}$$

Similarly, in the case for which

$$F = G+H = G+KL \tag{8.8–25}$$

but

$K+L \neq 1$

since $KL = K(\bar{K}+L)$, we have

$$F = G+K(\bar{K}+L) = G \# K \# (\bar{K}+L) \qquad (8.8–26$$

Now Equation (8.8–24) or (8.8–26) can be applied to an arbitrary function. Consider the form of Equation (8.8–17) and apply Equation (8.8–24). For $A = x_1 + F_{\bar{x}_1}$, $C = \bar{x}_1$, $D = F_{x_1}$, we have

$$\begin{aligned} F &= (x_1 + F_{\bar{x}_1})(\bar{x}_1 + x_1 F_{x_1}) \\ &= (x_1 + F_{\bar{x}_1}) \# \bar{x}_1 \# x_1 F_{x_1} \end{aligned} \qquad (8.8\text{–}27)$$

For $A = \bar{x}_1 + F_{x_1}$, $C = x_1$, $D = F_{\bar{x}_1}$, we have

$$F = (\bar{x}_1 + F_{x_1}) \# x_1 \# \bar{x}_1 F_{\bar{x}_1} \qquad (8.8\text{–}28)$$

By applying Equation (8.8–26) to the form of Equation (8.8–11), we obtain exactly the same expressions as Equations (8.8–27) and (8.8–28).

It is seen that Equation (8.8–27) or (8.8–28) contains a sum as well as a product, and therefore is called a *hybrid expansion.*

In this way the given function F will be realized in successive levels of composite form, i.e. in both sum and product forms.

Although this method is not necessarily optimum so far as the number of elements is concerned, it results in a network with fairly few elements, and compares favourably with other methods. This method is systematic and is easily applied.

This method can be extended to the synthesis of networks composed of five-input majority gates or seven-input majority gates, which will not be discussed here.

8.9 REALIZATION OF SYMMETRIC SWITCHING FUNCTIONS WITH THRESHOLD LOGIC ELEMENTS

As defined in Section 5.1, a switching function F of n variables is a symmetric function if it remains invariant to all permutations of all the n variables.

It is relatively easy to realize symmetric functions, and certain special methods can be employed. The realization of a symmetric function with a network of threshold logic elements by using algebraic methods has been studied by Minnick[19] and Kautz,[14] and the realization by using graphical method has been studied by Sheng.[24]

In this section we shall discuss briefly the basic concepts associated with the realization of symmetric functions, without going to the details of the methods.

To realize a symmetric function with threshold logic elements, by Theorem 5.3.5, it is admissible to have the same weight for all the n variables. For minimal-integral realization, it is obvious that each weight should be unity. So we shall take the weight of each variable to be unity, i.e. $a_i = 1, i = 1, 2, \ldots, n$.

A symmetric function F of n variables $x_1, \ldots, x_n$ is represented as $F = S_{\{i, j, \ldots, m\}}\{x_1, \ldots, x_n)$

iff it contains exactly all the n-tuples where the number of uncomplemented input variables is $i, j, \ldots,$ or m.

For example,

$$S_{\{1,4\}}(x_1, x_2, x_3, x_4)$$
$$= x_1\bar{x}_2\bar{x}_3\bar{x}_4 + \bar{x}_1x_2\bar{x}_3\bar{x}_4 + \bar{x}_1\bar{x}_2x_3\bar{x}_4 + \bar{x}_1\bar{x}_2\bar{x}_3x_4 + x_1x_2x_3x_4$$

is a symmetric function, symmetric in x_1, x_2, x_3, and x_4.

It is to be noted that an input variable itself may be in the complemented form already. Thus

$$S_{\{2\}}(x_1, \bar{x}_2, \bar{x}_3) = x_1\bar{x}_2x_3 + x_1x_2\bar{x}_3 + \bar{x}_1\bar{x}_2\bar{x}_3$$

is a symmetric function too, symmetric in x_1, $\bar{x}_2$, and $\bar{x}_3$, but not symmetric in x_1, x_2, and x_3.

The symmetric function $F = S_{\{i\}}(x_1, \ldots, x_n)$, $1 \leq i \leq n$, is called an *elementary symmetric function*, or a *σ-function*, and is designated as σ_i[1].

The symmetric function $F = S_{\{m, m+1, \ldots, n\}}(x_1, \ldots, x_n)$ $1 \leq m \leq n$, is called a *δ-function*, and is designated as δ_m^1.

It can be easily shown that a δ-function δ_m of n variables is realizable with a single threshold logic element with a threshold value m. When n is odd and $m = (n+1)/2$, the threshold logic element realizing the δ-function is an n-input majority gate.

The σ-function $S_{\{i\}}(x_1, \ldots, x_n)$, $i \neq 1$ or n, is not realizable with a single threshold logic element. The σ-function $S_{\{n\}}(x_1, \ldots, x_n)$ is obviously also a δ-function with $m = n$, and therefore is realizable with a single threshold logic element.

An input with a negative weight is called an *inhibition*, because when this input is present and if the negative weight is very large, this input alone will inhibit the output of the element from being 1, whatever the input combination of other input variables may be.

A σ-function $\sigma_m = S_{\{m\}}(x_1, \ldots, x_n)$ can be realized with a threshold logic element realizing the δ-function δ_m inhibited by another δ-function δ_{m+1} with a large negative weight. For instance, $S_{\{3\}}(x_1, \ldots, x_4)$ can be realized as $S_{\{3,4\}}(x_1, \ldots, x_4)$ inhibited by $S_{\{4\}}(x_1, \ldots, x_4)$ with a negative weight-2, as shown in Fig. 22.

It is obvious that each σ-function can be realized with two threshold logic elements. Consequently, an arbitrary function, which is the sum of j σ-functions, can be realized with at most $2j$ threshold logic elements.

However, $2j$ is a very loose upper bound of the number of threshold logic elements required to realize an arbitrary symmetric function. In fact, an arbitrary symmetric function can often be realized with much fewer threshold logic elements by inhibiting an element realizing a δ-function with other δ-functions.

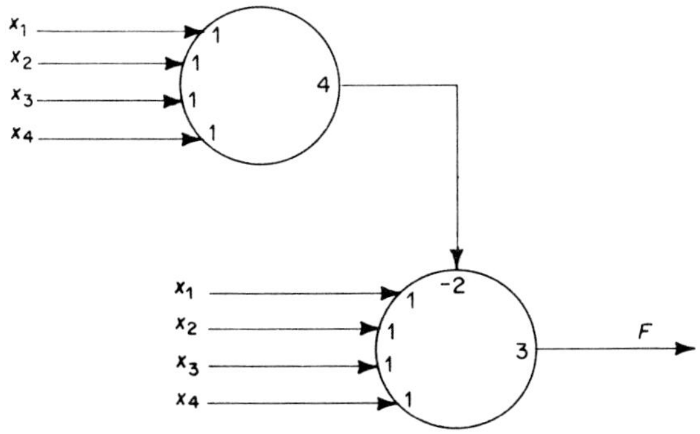

FIG. 22. Realization of $F = S_{\{3\}}(x_1, x_2, x_3, x_4) = \delta_3(x_1, x_2, x_3, x_4)$.

For instance, consider $F = S_{\{2,4,5,8\}}(x_1, \ldots, x_8)$. It can be realized as δ_2 inhibited by δ_3 with a negative weight-2, and also by δ_6 with a negative weight-4, as shown in Fig. 23.

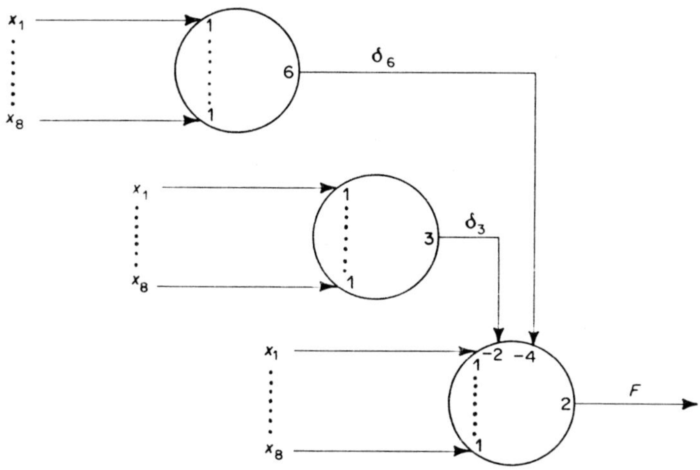

FIG. 23. Realization of $F = S_{\{2,4,5,8\}}(x_1, \ldots, x_8)$.

It is to be noted that, in conformity with the general structure of threshold-logic networks, element 1 in the above example could be inhibited by the output of element 2, or vice versa. Then the number of threshold logic elements required could be further reduced. The problem of compound synthesis for the realization of symmetric functions is to find a network, in conformity with the general structure, to realize the given function such that the total number of elements used is minimum or close to minimum. The details of the existing methods are all too lengthy to justify their being included here.

REFERENCES

1. Arnold, R. E., and Harrison, M. A. (1963). Algebraic properties of symmetric and partially symmetric Boolean functions, *I.E.E.E. Trans. electronic Comput.* EC**12,** 244–251.
2. Akers, S. B. (1962). Synthesis of combinational logic using three-input majority gates, in "Switching Circuit Theory and Logical Design", *A.I.E.E. Special Publication* S**134,** 149–157.
3. Akers, S. B. (1961). On the algebraic manipulation of majority logic, *I.R.E. Trans. electronic Comput.* EC**10,** 779.
4. Amarel, S., Cooke, G., and Winder, R. O. (1964). Majority gate networks, *I.E.E.E. Trans. electronic Comput.* EC**13,** 4–13.
5. Cameron, S. H. (1964). The generation of minimal threshold nets by an integer program, *I.E.E.E. Trans. electronic Comput.* EC**13,** 299–302.
6. Coates, C. L., and Lewis, P. M. II (1963). A realization procedure for threshold gate networks, *I.E.E.E. Trans. electronic Comput.* EC**12,** 454–461.
7. Cohn, M., and Lindaman, R. (1961). Axiomatic majority-decision logic, *I.R.E. Trans electronic Comput.* EC**10,** 17–21.
8. Fischler, M. A. (1962). Investigations concerning the theory and synthesis of linearly separable switching functions, Ph.D. dissertation, Department of Electrical Engineering, Stanford University, U.S.A.
9. Hanson, W. H. (1963). Threshold-logic synthesis by algebraic methods, *I.E.E.E. Trans. electronic Comput.* EC**12,** 401–402.
10. Hopcroft, J. E., and Mattson, R. L. (1965). Synthesis of minimal threshold logic networks, *I.E.E.E. Trans. electronic Comput.* EC**14,** 552–560.
11. Horna, O. A. (1963). Majority-logic synthesis, *I.E.E.E. Trans. electronic Comput.* EC**12,** 131–132.
12. Horna, O. A. (1965). A geometric method for three-input majority logic networks, *I.E.E.E. Trans. electronic Comput.* EC**14,** 475–481.
13. Hu, S. T. (1965). "Threshold Logic", University of California Press, U.S.A.
14. Kautz, W. H. (1961). The realization of symmetric switching functions with linear-input logic elements, *I.R.E. Trans. electronic Comput.* EC**10,** 371–378.
15. Lewis, P. M. II, and Coates, C. L. (1963). Realization of logical functions by a network of threshold components with specified sensitivity, *I.E.E.E. Trans. electronic Comput.* EC**12,** 443–454.
16. Lindaman, R. (1960). A theorem for deriving majority-logic networks within an augmented Boolean algebra, *I.R.E. Trans. electronic Comput.* EC**9,** 338–342.
17. Mattson, R. L. (1962). The analysis and synthesis of adaptive systems which use networks of threshold elements, *Technical Report No.* 1553–1, Solid-state Electronics Laboratory, Stanford University Laboratories.
18. Miller, H. S., and Winder, R. O. (1962). Majority-logic synthesis by geometric methods, *I.R.E. Trans. electronic Comput.* EC**11,** 89–90.
19. Minnick, R. C. (1961). Linear-input logic, *I.R.E. Trans. electronic Comput.* EC**10,** 6–16.
20. Miyata, F. (1963). Realization of arbitrary logical functions using majority elements, *I.E.E.E. Trans. electronic Comput.* EC**12,** 183–191.
21. Miyata, F. (1964). An extension of the method of Cohn and Lindaman, *I.E.E.E. Trans. electronic Comput.* EC**13,** 625–629.
22. Muroga, S., Toda, I., and Takasu, S. (1961). Theory of majority decision elements, *J. Franklin Inst.* **271,** 376–418.

23. Negrin, A. E. (1964). Synthesis of practical three-input majority logic networks, *I.E.E.E. Trans. electronic Comput.* EC**13,** 296–299.
24. Sheng, C. L. (1965). A graphical interpretation of realization of symmetric Boolean functions with threshold logic elements, *I.E.E.E. Trans. electronic Comput.* EC**14,** 8–18.
25. Sheng, C. L. (1965). Compound synthesis of threshold network for the realization of general Boolean functions. *I.E.E.E. Trans. electronic Comput.* EC**14,** 794–814.
26. Stram, O. B. (1961). Arbitrary Boolean functions of *n*-variables realizable in terms of threshold devices, *Proc. Inst. Radio Engrs* **49,** 210–220.
27. Tohma, Y. (1964). Decomposition of logical functions using majority decision elements, *I.E.E.E. Trans. electronic Comput.* EC**13,** 698–705.
28. Winder, R. O. (1962). Threshold Logic, Ph.D. dissertation, Department of Mathematics, Princeton University. U.S.A.

CHAPTER 9

Sensitivity of Weights and Threshold

9.1 INTRODUCTION

The realization problem studied in Chapters 6 and 7 is purely theoretical, in that if a switching function of n variables is a threshold function, then it is considered as realizable by a threshold logic element irrespective of the number of input variables n and the size of weights. In this chapter we shall study the *sensitivity* problem, which is closely related to physical circuit realization. Before presenting the sensitivity problem, it may be in order to have a brief discussion of the physical realization of threshold logic elements.

In physical realization, a threshold logic element can be considered as the cascading of two components. The first component is a summation device whose output is a sum of the quantities representing the input variables. The second component is a comparison device which compares the input with a threshold and gives an output as a nonlinear step function of the input. The symbolic diagram of such a cascading is shown in Fig. 24.

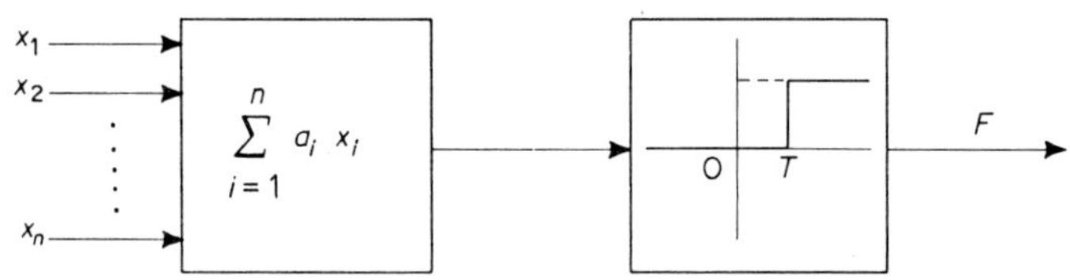

FIG. 24. Symbolic representation of a threshold logic element as a cascading of a summation device and a comparison device.

Physically a threshold logic element can be mechanized as several resistors feeding into an active component such as a vacuum tube, a transistor, a tunnel diode, etc. To give different weights to the input variables, the resistances may have different values, as contrary to the situations with AND and OR gates, where the resistances are all equal so as to give equal weights to all input variables. Then the active component is fired or not depending on whether the threshold value is attained or not.

For accurate summation of input quantities, say voltages, an operational amplifier can be used. Let the input voltages be either O or V volts,

$R_1, \ldots, R_n$ be the resistances in the input lines, and R_O be the feedback resistance. Then the output voltage is

$$V_O = -\left(\frac{R_O}{R_1} + \frac{R_O}{R_2} + \ldots + \frac{R_O}{R_n}\right) V$$

Thus the output voltage is equal to the weighted sum of the input voltages, with a change of sign. The weights $\{a_i\}$ are inversely proportional to the resistances in the input lines, $\{R_i\}$.

A difference amplifier made of, say, a two-transistor flip-flop, can also be used to mechanize a threshold logic element, as shown in Fig. 25.

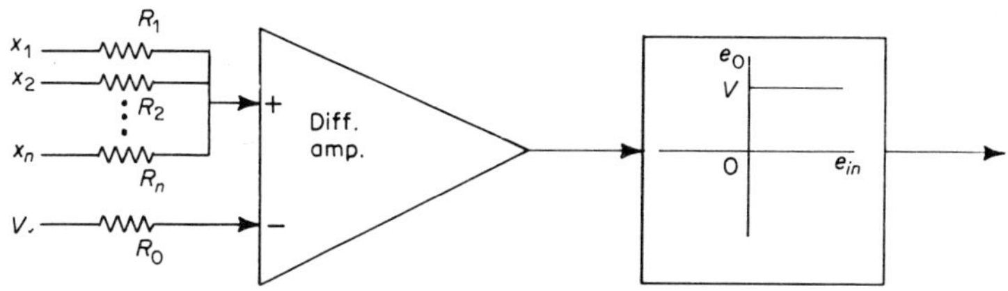

FIG. 25. Difference amplifier used as a threshold logic element.

The input variables $\{x_i\}$ are either O or V volts. The output of the difference amplifier is a positive or negative voltage depending on whether the current entering the + terminal exceeds the current entering the − terminal or not. Since the input impedance of either terminal is assumed to be very low and negligible, the current in each input line is inversely proportional to the respective resistance. Thus a_i is inversely proportional to R_i. In Fig. 25, each input variable is assumed to have a positive weight. If an input variable x_j is to have a negative weight, then R_j should be connected to the − terminal instead.

Threshold logic elements can also be mechanized by magnetic cores with transformer windings. Weighting can be obtained by using different numbers of turns for different input variables, because the magnetizing force is proportional to the number of turns. Since the binary values are usually represented by the polarity of the signals, $\{-1, 1\}$ instead of $\{0, 1\}$ should be used for the binary values.

Since in a physical realization input values and weights are represented by physical quantities, naturally they have variations. When the overall variation exceeds the gap length g, the output of a threshold logic element will be in error. The overall variation increases with the number of input variables and the size of weights. Thus practically there is a limit to the number of input variables and to the sizes of weights. For large n, sometimes it is desirable or even nccessary to decompose a function F so as to realize it with a number of threshold logic elements, even if F is 1-realizable.

The criterion for determining the allowable variation is sensitivity, or tolerance.

The sensitivity problem has been studied by Winder,[6] Wood,[7] Lewis and Coates,[4] Coates and Lewis,[2] Hu,[3] Blomgren and Torng,[1] Sheng,[5] etc. Usually it is studied in conjunction with the realization problem. However, since a non-threshold function is not 1-realizable for any sensitivity requirement, and a threshold function must be 1-realizable for some sensitivity requirement, we think the sensitivity problem comes after the realization problem, and therefore can be detached from the latter. We shall divide the problem into several sub-problems. First, for a given assignment of weights of a given threshold function, what are the maximum allowable threshold and weight sensitivities with which the function can be realized, and what is the threshold value corresponding to these maximum allowable sensitivities? Next, what are the factors that may affect the maximum allowable sensitivities, and how can they be utilized to increase the maximum allowable sensitivities? And finally, for a given threshold function, what is the realization that will result in the highest maximum allowable sensitivities? The last problem is not completely solved in general, but can be approximated by trial and error.

Before the analysis of the problem, we shall have a discussion of the definitions of sensitivities, in order to clarify certain assumptions and to justify their adoption.

9.2 DISCUSSION OF DEFINITIONS

The *sensitivity* of a single physical parameter a is defined as the ratio of the absolute value of the *maximum expected variation* in a, v_a, to the absolute value of the parameter itself, i.e. $v_a/|a|$.

Let

T be the threshold value.

v_T be the maximum expected variation in T, considered positive.

a_i be the weight of input x_i.

v_{a_i} be the maximum expected variation in a_i, considered positive.

Then

$$s_T = \frac{v_T}{|T|}, \text{ the } \textit{threshold sensitivity}.$$

$$s_{a_i} = \frac{v_{a_i}}{|a_i|}, \text{ the } \textit{weight sensitivity} \text{ of weight } a_i.$$

Now we want to find the *overall maximum expected variation* of the sum of several parameters and to define the *overall sensitivity*. Let the parameters be $a_1, a_2, \ldots, a_n$, and their maximum expected variations be $v_{a_1}, v_{a_2}, \ldots, v_{a_n}$, respectively.

First we assume that

$$\frac{v_{a_1}}{|a_1|} = \frac{v_{a_2}}{|a_2|} = \ldots = \frac{v_{a_n}}{|a_n|}$$

or

$$s_{a_1} = s_{a_2} = \ldots = s_{a_n} = s_a \tag{9.2–1}$$

This assumption is on a "constant percentage" basis. It is not necessarily true, because a larger parameter can be made more accurate percentagewise than a smaller parameter. However, because of lack of better choice and for the sake of simplicity, we adopt this assumption.

Next, according to convention, we assume that v_a, the overall maximum expected variation, is equal to the sum of all individual variations, i.e.

$$v_a = v_{a_1} + v_{a_2} + \ldots + v_{a_n} = \sum_{i=1}^{n} v_{a_i} \tag{9.2–2}$$

This assumption is rather controversial. If v_{a_i} is a definite quantity, then Equation (9.2–2) is of course true. However, the variation of a physical parameter is a random variable, whose distribution and maximum value are not exactly known. Thus v_{a_i} should be considered as an arbitrarily chosen value, say $k\sigma_i$, where k is a positive constant and σ_i is the standard deviation, such that the probability of the variation greater than $k\sigma_i$ or smaller than $-k\sigma_i$ is negligible. Assume that the variations in $a_1, a_2, \ldots, a_n$ all have the same distribution and therefore all are truncated at the same value of k. Then $\sum_{i=1}^{n} a_i$ will have a deviation $\sigma = \sqrt{\sigma_1^2 + \ldots + \sigma_n^2}$. If the overall variation is also truncated at $k\sigma$, then

$$\begin{aligned} v_a &= k\sigma \\ &= k\sqrt{\sigma_1^2 + \ldots + \sigma_n^2} \\ &= \sqrt{(k\sigma_1)^2 + \ldots + (k\sigma_n)^2} \\ &= \sqrt{v_{a_1}{}^2 + \ldots + v_{a_n}{}^2} \end{aligned} \tag{9.2–3}$$

Since the exact distribution of variations in $\{a_i\}$ is not known, the adoption of Equation (9.2–2) is not necessarily justified. However, since Equation (9.2–2) is on the safe side, and to conform with convention and for the sake of simplicity, we shall still adopt it. Thus we have

$$\begin{aligned} v_a &= v_{a_1} + v_{a_2} + \ldots + v_{a_n} \\ &= s_a|a_1| + \ldots + s_a|a_n| \\ &= s_a \sum_{i=1}^{n} |a_i| \end{aligned}$$

or

$$s_a = \frac{v_a}{\sum_{i=1}^{n} |a_i|} \tag{9.2–4}$$

Now s_a, the sensitivity of each individual weight, can also be considered as the overall sensitivity of the sum of m weights, and Equation (9.2–4) can be considered as the definition.

9.3 THE BASIC EQUATIONS

To study sensitivities, we are essentially dealing with the gap between the mapping of vertices in F and that of vertices in $\bar{F}$.

According to the definition of a threshold function, we have

$$F(x_1, \ldots, x_n) = 1 \quad \text{if} \quad f_A(x_1, \ldots, x_n) = A \,.\, X^t = \sum_{i=1}^{n} a_i x_i \geq T_u \tag{9.3–1}$$

$$F(x_1, \ldots, x_n) = 0 \quad \text{if} \quad f_A(x_1, \ldots, x_n) = A \,.\, X^t = \sum_{i=1}^{n} a_i x_i \leq T_l$$

where T_u is the upper boundary of the gap.

T_l is the lower boundary of the gap.

The threshold value T may be anywhere in the gap, or

$$T_l < T \leq T_u \tag{9.3–2}$$

And the gap length is

$$g = T_u - T_l \tag{9.3–3}$$

It is well known that for a given realization of a given threshold function, if T is replaced by $T' = T + \Delta T$, together with the addition of a constant input $x_0 = 1$ of weight ΔT, the function realized remains the same. When $\Delta T = -T$, $T' = 0$, and the threshold is completely replaced by a constant input. This kind of change sometimes may have effects on the sensitivities and will be studied later. At present we assume that there is no constant input x_0.

Let X_j, $j = 1, \ldots, m$, be a vertex in F.

X_k, $k = m+1, \ldots, 2^n$, be a vertex in $\bar{F}$.

$f_{A_j} = A \,.\, X_j^t$ be the value of f_A for X_j.

$f_{A_k} = A \,.\, X_k^t$ be the value of f_A for X_k.

$f_{|A_j|}$ be the sum of the absolute values of weights for X_j.

$f_{|A_k|}$ be the sum of the absolute values of weights for X_k.

v_{A_j} be the overall maximum expected variation in the weights for X_j.

v_{A_k} be the overall maximum expected variation in the weights for X_k.

Then we have the following basic relationships.

$$f_{A_j} - v_{A_j} \geq T + v_T, \quad j = 1, \ldots, m \tag{9.3–4}$$

$$T - v_T \geq f_{A_k} + v_{A_k}, \quad k = m+1, \ldots, 2^n \tag{9.3–5}$$

Since $v_T = s_T|T|$, $v_{A_j} = s_a f_{|A_j|}$, and $v_{A_k} = s_a f_{|A_k|}$, Inequalities (9.3–4) and (9.3–5) become

$$f_{A_j} - s_a f_{|A_j|} \geq T + s_T|T| \tag{9.3–6}$$

$$T - s_T|T| \geq f_{A_k} + s_a f_{|A_k|} \tag{9.3–7}$$

Rigorously all the 2^n inequalities should be satisfied. In most cases we can find the solution by solving only 2 equations with $(f_{A_j} - s_a f_{|A_j|})_{min}$ and $(f_{A_k} + s_a f_{|A_k|})_{max}$ as follows:

$$(f_{A_j} - s_a f_{|A_j|})_{min} = T + s_T|T| \tag{9.3–8}$$

$$T - s_T|T| = (f_{A_k} + s_a f_{|A_k|})_{max} \tag{9.3–9}$$

Equations (9.3–8) and (9.3–9) are called the *basic equations*, as all the development in the following are based on these two equations.

Thus the problem reduces to how to choose the minimum and maximum quantities. The problem is complicated by the fact that these minimum and maximum quantities are themselves functions of s_a, the weight sensitivity to be determined.

9.4 ANALYSIS OF THE BASIC EQUATIONS FOR VARIOUS CASES

The solution of the basic equations depends on the binary values chosen for x_i and also on the signs of the weights. We shall study the two most common cases, *viz.* $x_i \in \{-1, +1\}$ and $x_i \in \{0, 1\}$, separately.

CASE I: $x_i \in \{-1, +1\}$

In this case the weight a_i is always present, either as $+a_i$ or as $-a_i$, depending on whether x_i is $+1$ or -1. Therefore we have

$$f_{|A_j|} = f_{|A_k|} = \sum_{i=1}^{n} |a_i| = \text{constant}$$

$$(f_{A_j} - s_a f_{|A_j|})_{min} = f_{A_j\,(min)} - s_a \sum_{i=1}^{n} |a_i|$$

$$= T_u - s_a \sum_{i=1}^{n} |a_i|$$

$$(f_{A_k} + s_a f_{|A_k|})_{max} = f_{A_k(max)} + s_a \sum_{i=1}^{n} |a_i|$$

$$= T_l + s_a \sum_{i=1}^{n} |a_i|$$

Substituting into Equations (9.3–8) and (9.3–9), we have

$$T_u - s_a \sum_{i=1}^{n} |a_i| = T + s_T|T| \tag{9.4–1}$$

$$T - s_T|T| = T_l + s_a \sum_{i=1}^{n} |a_i| \tag{9.4–2}$$

For a given assignment of weights, T_u, T_l, and $\sum_{i=1}^{n} |a_i|$ are known. Since in these two equations there are three unknowns T, s_T, and s_a to be solved for, there exist infinite sets of solutions. If one of the three unknowns or another relationship among the unknowns is given, then the unknowns can be uniquely solved for, under the constraints that $T_l < T \leq T_u$, $s_T > 0$, and $s_a > 0$.

As a special case, consider $s_T = cs_a$, where c is a positive constant. Then Equations (9.4–1) and (9.4–2) reduce to

$$T_u - s_a \sum_{i=1}^{n} |a_i| = T + cs_a|T| \tag{9.4–3}$$

$$T - cs_a|T| = T_l + s_a \sum_{i=1}^{n} |a_i| \tag{9.4–4}$$

Solving, we have

$$T = \frac{T_u + T_l}{2} \tag{9.4–5}$$

$$s_a = \frac{T_u - T_l}{c|T_u + T_l| + 2\sum_{i=1}^{n} |a_i|} \tag{9.4–6}$$

CASE II: $x_i \in \{0, 1\}$

In this case there are again several different situations. One is that the weights are all positive and consequently the threshold value is also positive. (A negative threshold value implies a constant function $F = 1$.) A second situation is that some weights are positive and some are negative (mixed weights), and the threshold value is either positive or negative. A third situation is that the weights are all negative and consequently the threshold value is also negative. Since this last situation is similar to the first one, it will not be considered.

For $x_i \in \{0, 1\}$, we consider the "*perfect-zero model*",[4] in which the logical "0" corresponds to a zero-valued signal which does not vary.

CASE II–1: All positive weights.

Let $\{x_{i_1}, \ldots, x_{i_m}\}$ be the set of input variables that have a value 1 in X_j. Then

$$\begin{aligned} v_{A_j} &= v_{A_{i_1}} + \ldots + v_{A_{i_m}} \\ &= s_a a_{i_1} + \ldots + s_a a_{i_m} \\ f_{A_j} &= a_{i_1} + \ldots + a_{i_m} \end{aligned}$$

Therefore

$$\begin{aligned} f_{A_j} - v_{A_j} &= (1 - s_a)(a_{i_1} + \ldots + a_{i_m}) \\ &= (1 - s_a) f_{A_j} \end{aligned}$$

$$(f_{A_j} - v_{A_j})_{\min} = (1 - s_a) f_{A_j(\min)} = (1 - s_a) T_u$$

Similarly,

$$(f_{A_k} + v_{A_k})_{\max} = (1 + s_a) f_{A_k(\max)} = (1 + s_a) T_l$$

Thus Equations (9.3–8) and (9.3–9) reduce to

$$(1 - s_a) T_u = T + s_T |T| \tag{9.4–7}$$

$$T - s_T |T| = (1 + s_a) T_l \tag{9.4–8}$$

As a special case, when $s_T = s_a = s$, since $T > 0$, we have

$$T = \sqrt{T_u T_l} \tag{9.4–9}$$

$$s = \frac{T_u + T_l - 2\sqrt{T_u T_l}}{T_u - T_l} \tag{9.4–10}$$

CASE II–2: Mixed weights

Let $\{x_{i_1}, \ldots, x_{i_m}\}$ be the set of input variables that have a value 1 in X_j. Then

$$f_{A_j} = a_{i_1} + \ldots + a_{i_m}$$

$$f_{|A_j|} = \sum_{q=1}^{m} |a_{i_q}| \geq f_{A_j}$$

Similarly,

$$f_{|A_k|} \geq f_{A_k}$$

In this case the determination of $(f_{A_j} - s_a f_{|A_j|})_{\min}$ and $(f_{A_k} + s_a f_{|A_k|})_{\max}$ is not straightforward, because $(f_{A_j} - s_a f_{|A_j|})_{\min}$ may be obtained from a smaller f_{A_j} and a smaller $f_{|A_j|}$, or from a larger f_{A_j} and a larger $f_{|A_j|}$. Similarly $(f_{A_k} + s_a f_{|A_k|})_{\max}$ may be obtained from a smaller f_{A_k} and a larger $f_{|A_k|}$, or from a larger f_{A_k} and a smaller $f_{|A_k|}$. In other words, for each pair of $\{X_j, X_k\}$, the quantities f_{A_j}, $f_{|A_j|}$, f_{A_k}, and $f_{|A_k|}$ impose restrictions on the maximum allowable values of s_T and s_a, and the most severe restrictions are not necessarily imposed by $f_{A_j} = T_u$ and $f_{A_k} = T_l$. It is not known in advance which values of f_{A_j} and f_{A_k} will impose the most severe restrictions. Therefore in general a search process is required. If in Equations (9.3–8) and (9.3–9) certain values of f_{A_j} and f_{A_k} are used and s_T, s_a, and T are found, it is usually necessary to check the inequalities with other values of f_{A_j} and f_{A_k} for consistency.

A relatively simple situation is that $(f_{A_j} - s_a f_{|A_j|})_{\min}$ is obtained from $f_{A_j(\min)} = T_u$ and $(f_{A_k} + s_a f_{|A_k|})_{\max}$ is obtained from $f_{A_k(\max)} = T_l$. Then

$$(f_{A_j} - s_a f_{|A_j|})_{\min} = T_u - s_a f_{|A_j|\max}$$

$$(f_{A_k}+s_a f_{|A_k|})_{\max} = T+s_a f_{|A_k|\max}$$

where $f_{|A_j|\max}$ and $f_{|A_k|\max}$ can be found from the weights corresponding to vertices $\{X_j\}$ and $\{X_k\}$ such that $f_{A_j} = T_u$ and $f_{A_k} = T_l$. Then Equations (9.3–8) and (9.3–9) reduce to

$$T_u - s_a f_{|A_j|\max} = T+s_T|T| \qquad (9.4\text{–}11)$$

$$T - s_T|T| = T+s_a f_{|A_k|\max} \qquad (9.4\text{–}12)$$

EXAMPLE 9.4–1

$F(x_1, x_2, x_3, x_4) = x_1x_2+x_1x_3+x_2x_3x_4$

$x_i \in \{0, 1\}$

$R[F] = (3, 2, 2, 1; T)$

$T_u = 5, T_l = 4, \qquad 4 < T \leq 5$

Assume that $s_T = s_a = s$

From Equations (9.4–9) and (9.4–10), we have

$T = 4.472$

$s = 0.056.$

9.5 EFFECT OF CHANGE OF SIZES OF WEIGHTS

Since different assignments of weights of the same threshold function may result in different *maximum allowable sensitivities*, we shall study the effects of change of weights. There are three different kinds of changes. One is the change of sizes of weights; another is the addition of a constant input ΔT with a corresponding change in threshold value; the third is the change of signs of a subset of the weights.

The effect of change of sizes of weights is rather involved for mixed weights. The highest maximum allowable sensitivities may be obtained from a non-minimal-integral assignment of mixed weights. Here we shall study its effects on realizations of all positive weights only.

THEOREM 9.5.1. For a given assignment of weights of a given threshold function, if every weight and the threshold value are multiplied by the same constant k, the maximum allowable sensitivities remain the same.

PROOF: This is obvious from the basic equations. That every weight and the threshold value are multiplied by a constant k is equivalent to that every term in Equations (9.3–8) and (9.3–9) is multiplied by a constant k, which can be cancelled out later. Thus the solution remains the same.

Q.E.D.

Now we can limit our study to the case of normal realization, i.e. $g = 1$. When all weights are positive, obviously the minimal assignment will result in the highest maximum allowable sensitivities. In most cases such a minimal assignment is also integral. As pointed out by Winder[6] and Hu,[3] it is only in certain very unusual cases that the minimal assignment is non-integral. Therefore it is well justified that we consider minimal-integral assignment of weights only. Thus we have the following theorem.

THEOREM 9.5.2. For a threshold function realized with all positive weights, the minimal-integral assignment of weights gives the highest maximum allowable sensitivities.

9.6 EFFECT OF A CONSTANT INPUT $x_0 = 1$

It is obvious that if we add a fixed amount ΔT to the threshold value and also to the weight of a constant input $x_0 = 1$, the threshold function realized remains the same.

Consider Equations (9.3–8) and (9.3–9). It is easily seen that as long as $s_T > 0$ it is always possible to increase s_a by decreasing s_T, and vice versa. Therefore to study the effect of adding ΔT, we must assume a constant ratio between s_T and s_a, or let $s_T = cs_a$, where $c > 0$. Consider X_j and X_k as the vertices for $(f_{A_j} - s_a f_{|A_j|})_{\min}$ and $(f_{A_k} + s_a f_{|A_k|})_{\max}$. Then Equations (9.3–8) and (9.3–9) become

$$f_{A_j} - s_a f_{|A_j|} = T + cs_a|T| \qquad (9.6\text{–}1)$$

$$T - cs_a|T| = f_{A_k} + s_a f_{|A_k|} \qquad (9.6\text{–}2)$$

Adding Equations (9.6–1) and (9.6–2) and rearranging, we have

$$s_a(2c|T| + f_{|A_j|} + f_{|A_k|}) = f_{A_j} - f_{A_k} \qquad (9.6\text{–}3)$$

Now there are four different cases: (1) T and ΔT are both positive; (2) T is positive and ΔT is negative; (3) T and ΔT are both negative; (4) T is negative and ΔT is positive.

Consider case (1) first, i.e. both T and ΔT are positive. After adding ΔT, it is obvious that f_{A_j}, $f_{|A_j|}$, f_{A_k}, $f_{|A_k|}$, and T are each increased by ΔT. Let s'_a be the new maximum allowable weight sensitivity. Then we have

$$\begin{aligned} s'_a[2c(T+\Delta T) &+ (f_{|A_j|} + \Delta T) + (f_{|A_k|} + \Delta T)] \\ &= (f_{A_j} + \Delta T) - (f_{A_k} + \Delta T) \\ &= f_{A_j} - f_{A_k} \qquad (9.6\text{–}4) \end{aligned}$$

Equating Equations (9.6–3) and (9.6–4) and rearranging, we have

$$(s_a - s'_a)(2cT + f_{|A_j|} + f_{|A_k|}) = 2s'_a(c+1)\Delta T > 0$$

Since c, T, ΔT, $f_{|A_j|}$, and $f_{|A_k|}$ are all positive,

$$s_a - s'_a > 0$$

or

$$s_a > s'_a$$

It is seen that the new sensitivity is always smaller than the original one, or there is no improvement in sensitivity.

Next consider case (2), i.e. T is positive and ΔT is negative. After adding

ΔT, it is obvious that f_{A_j}, f_{A_k}, and T are increased by a negative quantity ΔT, while $f_{|A_j|}$ and $f_{|A_k|}$ are increased by a positive quantity $-\Delta T$. Assume that $|T| > |\Delta T|$, or $T+\Delta T > 0$. Then we have

$$\begin{aligned} & s_a'[2c(T+\Delta T)+f_{|A_j|}-\Delta T+f_{|A_k|}-\Delta T] \\ & \quad = f_{A_j}+\Delta T-(f_{A_k}+\Delta T) \\ & \quad = f_{A_j}-f_{A_k} \end{aligned} \tag{9.6–5}$$

Equating Equations (9.6–3) and (9.6–5) and rearranging, we have

$$(s_a-s_a')(2cT+f_{|A_j|}+f_{|A_k|}) = 2s_a'(c-1)\Delta T$$

The second factor on the left-hand side is always positive. Since ΔT is negative, the right-hand side is positive when $c < 1$, zero when $c = 1$, and negative when $c > 1$. Therefore we have

$$\begin{aligned} s_a &< s_a' \qquad \text{when } c > 1 \\ s_a &= s_a' \qquad \text{when } c = 1 \\ s_a &> s_a' \qquad \text{when } c < 1 \end{aligned}$$

The result of case (3) is the same as case (1), and that of case (4) is the same as case (2). Therefore we can establish the following theorem.

THEOREM 9.6.1. Let the threshold value and the weight of a constant input x_0 be both increased by an amount ΔT. If the absolute value of T increases, then the maximum allowable sensitivities decrease. If the absolute value of T decreases and $|T| > |\Delta T|$, then the maximum allowable sensitivities increase when $s_T > s_a$, remains the same when $s_T = s_a$, and decrease when $s_T < s_a$.

EXAMPLE 9.6–1

Consider F of Example 9.4–1 again. Assume that $s_T = 2s_a$.

Using Equations (9.4–7) and (9.4–8) we have

$5(1-s_a) = T+2s_aT$

$T-2s_aT = 4(1+s_a)$

Solving, we have

$T = 4.481$

$s_a = 0.037$

$s_T = 0.074$

Now let us introduce a constant input $x_0 = 1$ with a weight -4.

Then

$T_u' = T_u-4 = 5-4 = 1$

$T_l' = T_l-4 = 4-4 = 0$

$0 < T' \leq 1$

$f'_{|A_j|\max} = 5+|-4| = 9$

$f'_{|A_k|\max} = 4+|-4| = 8$

Using Equations (9.4–11) and (9.4–12), we have

$1-9s'_a = T'+2s'_a T'$

$T'-2s'_2 T' = 0+8s'_a$

Solving, we have

$T' = 0.0473$

$s'_a = 0.053$

$s'_T = 0.106$

It is to be noted that T' is solved from the equations, and is slightly different from $T-4 = 0.481$. However, if we took $T' = 0.481$, then s'_a and s'_T would be slightly smaller than the above calculated values, but still greater than the original ones.

9.7 EFFECT OF CHANGE OF SIGN OF A SUBSET OF WEIGHTS

By Corollary 5.1.1–1, for a threshold function $F(x_1, \ldots, x_n)$ with a realization $R[F] = (a_1, \ldots, a_n; T)$, if we change the signs of a subset of the weights, say $\{a_{i_1}, \ldots, a_{i_m}\}$, change the subset of input variables $\{x_{i_1}, \ldots, x_{i_m}\}$ into $\{\bar{x}_{i_1}, \ldots, \bar{x}_{i_m}\}$, and change T to $T-(a_{i_1}+\ldots+a_{i_m})$, then the threshold function realized remains the same.

Now we shall study the effect of such a change on the sensitivities.

The general case is very involved, and it is difficult to obtain definite conclusions. So we shall analyse the different cases separately.

First, consider Case 1 of Section 9.4, i.e. $x_i \in \{-1, +1\}$.

The realization of a threshold function with $x_i \in \{-1, +1\}$ can be obtained from the realization with $x_i \in \{0, 1\}$ by a transformation. Designate the variables after transformation $\{y_i\}$, the threshold T_y, and the weights $\{a_{i(y)}\}$. Then the transformation equations are:

$$y_i = 2x_i - 1, \; a_{i(y)} = a_i, \; T_y = 2T - \sum_{i=1}^{n} a_i$$

Let $a = a_{i_1}+\ldots+a_{i_m}$. After the change of signs of a subset of weights, we have

$T' = T-a$

$$\sum_{i=1}^{n} a'_i = \sum_{i=1}^{n} a_i - 2a$$

For $y_i \in \{-1, +1\}$,

$$a'_{i(y)} = a'_i$$

$$T'_y = 2T' - \sum_{i=1}^{n} a'_i = 2(T-a) - \left(\sum_{i=1}^{n} a_i - 2a\right)$$

$$= 2T - \sum_{i=1}^{n} a_i = T_y$$

It is seen that after the change of signs the threshold value remains the same. It can be easily shown that all f_{A_j} and f_{A_k}, and consequently T_u and T_l, remain the same after the change of signs of any subset of the weights. This implies that Equations (9.4–1) and (9.4–2) and the solution remain the same. Therefore we can establish the following theorem.

THEOREM 9.7.1. For $x_i \in \{-1, +1\}$, the maximum allowable sensitivities remain the same after the change of signs of any subset of the weights.

Next, consider Case II of Section 9.4, i.e. $x_i \in \{0, 1\}$.

For convenience, Equation (9.6–3) is reproduced here.

$$s_a(2c|T| + f_{|A_j|} + f_{|A_k|}) = f_{A_j} - f_{A_k} \tag{9.6–3}$$

Assume that either $T > 0$, $a > 0$, and $T - a \geq 0$, or $T < 0$, $a < 0$, and $T - a \leq 0$. Let $f'_{A_j} = f_{A_j} + \Delta f_{A_j}$, $f'_{A_k} = f_{A_k} + \Delta f_{A_k}$, $f'_{|A_j|} = f_{|A_j|} + \Delta f_{|A_j|}$, and $f'_{|A_k|} = f_{|A_k|} + \Delta f_{|A_k|}$ be the values of f_{A_j}, f_{A_k}, $f_{|A_j|}$, and $f_{|A_k|}$ after the change of signs of a subset of weights, where Δf_{A_j}, Δf_{A_k}, $\Delta f_{|A_j|}$, and $\Delta f_{|A_k|}$ are the increments. Also let s'_a be the maximum allowable sensitivity after change of signs. Then

$$s'_a(2c|T-a| + f'_{|A_j|} + f'_{|A_k|}) = f'_{A_j} - f'_{A_k}$$

or

$$\begin{aligned} s'_a[2c(|T| - |a|) + f_{|A_j|} + \Delta f_{|A_j|} + f_{|A_k|} + \Delta f_{|A_k|} \\ = (f_{A_j} - f_{A_k}) + (\Delta f_{A_j} - \Delta f_{A_k}) \end{aligned} \tag{9.7–1}$$

Combining Equations (9.6–3) and (9.7–1), we have

$$(s'_a - s_a)(2c|T| + f_{|A_j|} + f_{|A_k|}) = s'_a(2c|a| - \Delta f_{|A_j|} - \Delta f_{|A_k|}) + (\Delta f_{A_j} - \Delta f_{A_k}) \tag{9.7–2}$$

From Equation (9.7–2) it is seen that if the right-hand side is positive, then $s'_a > s_a$ and the sensitivity is increased. Now we shall study what is the condition for this to be true. After $x_{i_1}, \ldots, x_{i_m}$ are negated, the input vectors (in terms of the original variables) corresponding to $f'_{A_j} = f_{A_j} + \Delta f_{A_j}$ and $f'_{A_k} = f_{A_k} + \Delta f_{A_k}$ are the same as those corresponding to f_{A_j} and f_{A_k}, respectively. It is easily seen that

$$\Delta f_{A_j} = \Delta f_{A_k} = -a$$

or

$$\Delta f_{A_j} - \Delta f_{A_k} - 0 \tag{9.7–3}$$

Now let $\langle a \rangle = |a_{i_1}| + \ldots + |a_{i_m}|$ be the sum of absolute values of those weights whose signs have been changed. It is obvious that for every input vector, the change in the sum of absolute values of weights is at most $\langle a \rangle$. Therefore

$$\Delta f_{|A_j|} \leq \langle a \rangle, \qquad \Delta f_{|A_k|} \leq \langle a \rangle \tag{9.7–4}$$

Equations (9.7–3) and (9.7–4) are still not sufficient to make s'_a greater than s_a. So for mixed weights or an arbitrary value of c, there is no systematic way of improving s_a except by trial and error. However, when originally the weights are all positive or all negative, and $c \geq 1$, then

$$2c|a| \geq 2\langle a \rangle \geq \Delta f_{|A_j|} + \Delta f_{|A_k|}$$

or

$$2c|a| - \Delta f_{|A_j|} - \Delta f_{|A_k|} > 0 \qquad (9.7\text{–}5)$$

Now Equations (9.7–3) and (9.7–5) make the right-hand side of Equation (9.7–2) non-negative. Thus

$$s_a' - s_a \geq 0, \quad \text{or} \quad s_a' \geq s_a$$

It can be easily shown that when $T > 0$ and $a < 0$, or $T < 0$ and $a > 0$, then s_a tends to decrease. With these results, we can establish the following theorem.

THEOREM 9.7.2. For $x_i \in \{0, 1\}$, if initially the weights are all positive or negative, and $s_T \geq s_a$, the maximum allowable sensitivities can never be decreased, and sometimes can be increased, by the change of signs of any subset of weights, $\{a_{i_1}, \ldots, a_{i_m}\}$, provided that $|\sum_{q=1}^{m} a_{iq}| \leq |T|$.

9.8 REALIZATION METHOD AND EXAMPLE

Owing to the fact that Theorem 9.7.2 is not general enough, there seems to be no straightforward way of finding the optimum realization for an arbitrary threshold function, or a realization that will give the highest maximum allowable sensitivities. However, according to the theory developed above, we can find a realization fairly close to optimum. First we find the minimal integral assignment of weights. Set $a = T_l$, and find all the subsets of weights such that $a_{i_1} + \ldots + a_{i_m} = a$. Then choose a particular subset such that $f'_{|A_j|\max} + f'_{|A_k|\max}$ is the smallest. In finding this subset we can use a tabular form, as illustrated in the following example. Then set $T_u' = T_u - a$, $T_l' = T_l - a$, $f'_{|A_j|\max} = (f_{|A_j|} + \Delta f_{|A_j|})_{\max}$, and $f'_{|A_k|\max} = (f_{|A_k|} + \Delta f_{|A_k|})_{\max}$. Finally, by using Equations (9.4–11) and (9.4–12), together with the given relation between s_T and s_a, solve for T, s_T, and s_a.

This method, of course, does not guarantee the optimum solution, because with mixed weights, sometimes it is possible to obtain the highest allowable sensitivities from a non-minimal-integral assignment. Furthermore, the sensitivities thus obtained may even exceed the true maximum allowable sensitivities. This is because of the fact that $(f_{A_j} - s_a f_{|A_j|})_{\min}$ is not necessarily obtainable from $f_{A_j} = T_u$, and $(f_{A_k} + s_a f_{|A_k|})_{\max}$ is not necessarily obtainable from $f_{A_k} = T_l$. Thus, to ensure that the calculated T, s_T, and s_a are correct, it is necessary to check them with other inequalities for $f_{A_j} = T_u + 1$, $f_{A_j} = T_u + 2$, ..., and for $f_{A_k} = T - 1$, $T - 2$, ...

EXAMPLE 9.8–1

$F(x_1, \ldots, x_5) = x_1x_2 + x_1x_3 + x_1x_4 + x_2x_3 + x_2x_4x_5$

$x_i \in \{0, 1\}$

$R[F] = (5, 4, 3, 2, 1; T)$

$T_u = 7, T_l = 6, 6 < T \leq 7.$

Assume that $s_T = s_a = s$

From Equations (9.4–9) and (9.4–10), we have

$T = 6.481, \qquad s = 0.038$

Since the weights are all positive, according to Theorem 9.7.2, there is no harm in changing the signs of certain subsets of weights. So we shall try this to see if the maximum allowable sensitivities can be increased or not.

Set $a = T_l = 6$. Then the subsets of weights such that $\sum_{q=1}^{m} a_{i_q} = 6$ are $\{a_1, a_5\}, \{a_2, a_4\}, \{a_3, a_4, a_5\}$.

Let $\{X_j\}$ be the set of vertices such that $f_{A_j} = 7$, $\{X'_j\}$ be the set of vertices after a subset of variables has been negated, $\{X_k\}$ be the set of vertices such that $f_{A_k} = 6$, $\{X'_k\}$ be the set of vertices after a subset of variables has been negated, $f'_{|A_j|}$ be the sum of absolute values of weights corresponding to X'_j, and $f'_{|A_k|}$ be the sum of absolute values of weights corresponding to X'_k. These vertices and sum of weights for the three subsets of weights are listed in Table XVII.

TABLE XVII

	$\{a_1, a_5\}$		$\{a_2, a_4\}$		$\{a_3, a_4, a_5\}$	
X_j	X'_j	$f'_{\|A_j\|}$	X'_j	$f'_{\|A_j\|}$	X'_j	$f'_{\|A_j\|}$
10010	00011	3	11000	(9)	10101	(9)
01100	11101	(13)	00110	5	01011	7
01011	11010	11	00001	1	01100	7
X_k	X'_k	$f'_{\|A_k\|}$	X'_k	$f'_{\|A_k\|}$	X'_k	$f'_{\|A_k\|}$
10001	00000	0	11011	(12)	10110	(10)
01010	11011	(12)	00000	0	01101	8
00111	10110	10	01101	8	00000	0

For each subset of weights, $f'_{|A_j|\max}$ and $f'_{|A_k|\max}$ are circled.

For $\{a_1, a_5\}$, $f'_{|A_j|\max} + f'_{|A_k|\max} = 25$

For $\{a_2, a_4\}$, $f'_{|A_j|\max} + f'_{|A_k|\max} = 21$

For $\{a_3, a_4, a_5\}$, $f'_{|A_j|\max} + f'_{|A_k|\max} = 19$

Therefore we choose the last subset.

Now we have

$T'_u = T_u - a = 7-6 = 1$

$T'_l = T_l - a = 6-6 = 0$

$1 \geq T' > 0$

$f'_{|A_j|\max} = 9$

$f'_{|A_k|\max} = 10$

$s'_T = s'_a = s'$

Substituting these values into Equations (9.4–11) and (9.4–12), we have

$1-9s' = T'+s'T'$

$T'-s'T' = 0+10s'$

Solving, we have

$T' = 0.525$

$s' = 0.05$

After checking it is found that all inequalities are satisfied. Thus there is an increase in sensitivities over the original ones.

9.9 INCREASING SENSITIVITIES BY SPREADING f_{A_j} and f_{A_k}

From the table of Example 9.8–1 it is seen that for the same value of $f_{A_j} = T_u$ there are three X_j's. Similarly there are three X_k's for the same value of $f_{A_k} = T_l$. For functions of larger number of variables, the numbers of X_j's and X_k's for the same value of f_A will be even greater. This makes it difficult to have small values of $f'_{|A_j|\max}$ and $f'_{|A_k|\max}$ after the change of signs of a subset of weights. If we assign non-minimal integral values to the weights, sometimes it is possible to reduce the number of X_j's and X_k's corresponding to $f_{A_j} = T_u$ and $f_{A_k} = T_l$, or "spreading" f_{A_j} and f_{A_k} to several different values, respectively. Then it may be possible to find a subset of weights such that after the change of signs $(f'_{|A_j|\max} + f'_{|A_k|\max})$ will be smaller, thus resulting in higher values of maximum allowable sensitivities. This will be illustrated by the following example.

EXAMPLE 9.9–1

$F = x_1 + x_2x_3 + x_2x_4x_5$

$x_i \in \{0, 1\}$

$R[F] = (5, 3, 2, 1, 1; T)$

$T_u = 5, \quad T_l = 4, \quad 4 < T \leq 5$

Assume that $s_T = s_a = s$

From Equations (9.4–9) and (9.4–10), we have

$T = 4.472$

$s = 0.056$

It is found that if we set $a = 4$ and change the signs of any one of the subsets of weights $\{a_2, a_4\}$, $\{a_2, a_5\}$, $\{a_3, a_4, a_5\}$, we shall have

$$f'_{|A_j|\max} = 9 \quad \text{and} \quad f'_{|A_k|\max} = 6$$

Since $T'_u = 5-4 = 1$ and $T'_l = 4-4 = 0$, from Equations (9.4–11) and (9.4–12), we have

$$T' = 0.405$$
$$s' = 0.0633$$

It is seen that s is increased, but only slightly.

Now we shall try a non-minimal integral assignment. Let

$$R[F] = (7, 4, 2, 1, 1; T)$$
$$T_u = 6, \quad T_l = 5, \quad 5 < T \leq 6$$

With all positive weights, the maximum allowable sensitivity obtained is, of course, smaller than that obtained from the minimal integral assignment. However, if we choose $\{a_2\}$ as the subset of weights whose sign is to be changed, a much higher maximum allowable sensitivity can be obtained.

In using mixed weights, we have to examine the restrictions imposed by all values of f_{A_j} and f_{A_k}.

Now $a = 4$, $T'_u = 6-4 = 2$, and $T'_l = 5-4 = 1$. It is found that the most severe restrictions are imposed by $f'_{A_j} = T'_u + 1 = 3$ and $f'_{A_k} = T'_l = 1$.

The values of X_j, X'_j, $f'_{|A_j|}$, X_k, X'_k, and $f'_{|A_k|}$ for the subset $\{a_2\}$ are listed in Table XVIII.

TABLE XVIII

X_j	X'_j	$f'_{\vert A_j\vert}$
10000	11000	(11)
01110	00110	3
01101	00101	3
X_k	X'_k	$f'_{\vert A_k\vert}$
01010	00010	(1)
01001	00001	1

$f'_{|A_j|\max} = 11$

$f'_{|A_k|\max} = 1$

Substituting into Equations (9.3–8) and (9.3–9), we have

$3 - 11s' = T' + s'T'$

$T' - s'T' = 1 + s'$

Solving, we have

$T' = 1.317$

$s' = 0.137$

It is found that all the inequalities are satisfied, or the restrictions imposed by f'_{A_j} and f'_{A_k} of other values are not as severe as $f'_{A_j} = 3$ and $f'_{A_k} = 1$. Therefore the above solution is correct. However, as mentioned before, the value of $s = 0.137$ is not necessarily the highest maximum allowable sensitivity, considering all the possible integral assignments of weights and all the possible subsets of weights for change of signs.

REFERENCES

1. Blomgren, G. H., and Torng, H. C. (1966). Single threshold device realization subject to sensitivity requirements, *J. Franklin Inst.* **281,** 143–153.
2. Coates, C. L., and Lewis, P. M. II. (1963). A realization procedure for threshold gate networks, *I.E.E.E. Trans. electronic Comput.* EC**12,** 454–461.
3. Hu, S. T. (1965). "Threshold Logic", University of California Press, Berkeley and Los Angeles, U.S.A.
4. Lewis, P. M. II., and Coates, C. L. (1963). Realization of logical functions by a network of threshold components with specified sensitivity, *I.E.E.E. Trans. electronic Comput.* EC**12,** 443–454.
5. Sheng, C. L. (1966). On the satisfaction of sensitivity requirements by threshold logic elements, *Technical Report No.* 66–10, Department of Electrical Engineering, University of Ottawa.
6. Winder, R. O. (1962). Threshold Logic, doctoral dissertation, Mathematics Department, Princeton University, U.S.A.
7. Wood, P. E. Jr. (1963). A note on threshold device error analysis, *I.E.E.E. Trans. electronic Comput.* EC**12,** 403–405.

CHAPTER 10

Applications of Threshold Logic Elements

10.1 INTRODUCTION

In this chapter we shall study the applications of threshold logic elements. As will be shown in Section 10.2, a threshold logic element is a general gate, in the sense that it can be used as an AND gate, OR gate, NAND gate, NOR gate, or an inverter. Thus a threshold logic element can be used in any switching circuit where gates are required. However, since at the present time a threshold logic element is in general more costly than a conventional gate, it has not been as popularly used as conventional gates yet. It is expected that threshold logic elements will have a wider adoption in the future as the expense of physical realization decreases.

Since a threshold logic element can be used to replace any conventional gate, the number of threshold logic elements required to realize an arbitrary switching function is at most equal to that of conventional gates required to realize the same function, and is in most cases much smaller. It is in those situations in which the number of elements required is greatly reduced that the use of threshold logic elements is particularly advantageous and desirable.

A special type of threshold logic element, *viz.* an element with the same weight for all the inputs, is particularly important in that its physical implementation is relatively simple. It can be very conveniently used for the realization of symmetric switching functions. A further special type of this, the majority gate, is quite popularly used. An arbitrary switching function can be realized by a network of majority gates of the same type, for instance three-input majority gates, as was discussed in Chapter 8.

Although a threshold logic element itself is a combinatorial gate, like other types of gates, it can be used in a sequential circuit for the combinatorial logic part. Studies on the application of threshold logic elements to sequential machines have been started recently.[5, 6, 7]

So far there has been constructed at least one computer whose logic circuits are made exclusively of threshold logic elements.[1]

In Section 10.2 we shall show that the threshold logic element is a general gate. In Section 10.3 we shall give the realization of some typical arithmetic circuits by threshold logic elements. In Section 10.4 we shall illustrate how a threshold logic element can be used as an error-correcting device. In Section 10.5 we shall discuss adaptive or trainable threshold logic elements very briefly, and see how threshold logic elements can be conveniently used in learning machines, especially pattern classification.

Another very specific but interesting application of threshold logic elements is the transformation of probability, which will be studied in some detail in Chapter 11.

10.2 THRESHOLD LOGIC ELEMENT AS A GENERAL GATE

The ordinary AND, OR, NAND, and NOR gates each realize a special symmetric function which can also be realized by a threshold logic element. Thus a threshold logic element can be considered as a general gate. We shall study the four types of gates separately.

(1) AND gate

For an AND gate of n variables $\{x_1, x_2, \ldots, x_n\}$ the function realized is

$$\begin{aligned} F &= x_1\, x_2 \ldots x_n \\ &= S_{\{n\}}(x_1, x_2, \ldots, x_n) \end{aligned}$$

This is obviously a threshold function, for which the minimal-integral realization is

$$R[F] = (1, 1, \ldots; n)$$

(2) OR gate

For an OR gate of n variables $\{x_1, x_2, \ldots, x_n\}$, the function realized is

$$\begin{aligned} F &= x_1 + x_2 + \ldots + x_n \\ &= S_{\{1,\ldots,n\}}(x_1, x_2, \ldots, x_n) \end{aligned}$$

This is also a threshold function, for which the minimal integral realization is

$$R[F] = (1, 1, \ldots; 1)$$

(3) NAND gate

For a NAND gate of n variables $\{x_1, x_2, \ldots, x_n\}$, the function realized is

$$\begin{aligned} F &= \overline{(x_1 x_2 \ldots x_n)} \\ &= \bar{x}_1 + \bar{x}_2 + \ldots + \bar{x}_n \\ &= S_{\{0,\,1,\ldots,\,n-1\}}(x_1, x_2, \ldots, x_n) \\ &= S_{\{1,\,2,\ldots,\,n\}}(\bar{x}_1, \bar{x}_2, \ldots, \bar{x}_n) \end{aligned}$$

This is obviously also a threshold function. Let $F'(\bar{x}_1, \bar{x}_2, \ldots, \bar{x}_n) = F(x_1, x_2, \ldots, x_n)$ be F expressed as a function of the n negated variables $\{\bar{x}_1, \bar{x}_2, \ldots, \bar{x}_n\}$. Then the minimal-integral realization of F' is

$$R[F'] = (1, 1, \ldots, 1; 1)$$

By Corollary 5.1.1–1, we can realize F with a threshold logic element by complementing the input variables, changing the signs of the weights, and setting $T = T' - \sum_{i=1}^{n} a_i = 1-n$. Thus we have

$$R[F] = (-1, -1, \ldots, -1; 1-n)$$

(4) NOR gate

For a NOR gate of n variables $\{x_1, x_2, \ldots, x_n\}$, the function is

$$\begin{aligned} F &= \overline{(x_1+x_2+\ldots+x_n)} \\ &= \bar{x}_1\bar{x}_2 \ldots \bar{x}_n \\ &= S_{\{0\}}(x_1, x_2, \ldots, x_n) \\ &= S_{\{n\}}(\bar{x}_1, \bar{x}_2, \ldots, \bar{x}_n) \end{aligned}$$

This is also a threshold function. Let $F'(\bar{x}_1, \bar{x}_2, \ldots, \bar{x}_n) = F(x_1, x_2, \ldots, x_n)$ be F expressed as a function of the n negated variables $\{\bar{x}_1, \bar{x}_2, \ldots, \bar{x}_n\}$. Then the minimal-integral realization of F, is

$$R[F'] = (1, 1, \ldots, 1; n)$$

And the realization of F is

$$R[F] = (-1, -1, \ldots, -1; 0)$$

Trivially a threshold logic element can realize $\bar{x}_i$, the complementation of a variable x_i, and serves as an inverter. This is in fact a special case of either the NAND or the NOR gate. The minimal-integral realization is

$$R[F] = (-1; 0)$$

It is to be noted that a threshold logic element cannot realize the function realized by an exclusive OR gate, or, in general, cannot realize any linear function of two or more variables and also any elementary symmetric function $\sigma_{\{k\}}(x_1, x_2, \ldots, x_n)$ except for $k = 0$ and n.

10.3 REALIZATION OF CONVENTIONAL ARITHMETIC CIRCUITS BY THRESHOLD LOGIC ELEMENTS

The realization of conventional arithmetic circuits by threshold logic elements is extremely simple and economical because the number of elements required can be reduced to a minimum. Fischler,[2] and Fischler and Poe[3] have presented the circuits for some adders, subtractors, and complement generators. As an example the circuits for binary coded decimal adder, binary coded decimal subtractor and binary coded decimal complement generator are reproduced in Figs. 26, 27, and 28, respectively.

The double arrow symbolically designates the entire set of primary variables, each with a weight as given by its subscript. Primary inputs for Figs. 26 and 27 are (x_8, x_4, x_2, x_1), (y_8, y_4, y_2, y_1), and C_1, and have weights equal to their subscripts. Primary inputs for Fig. 28 are (x_8, x_4, x_2, x_1).

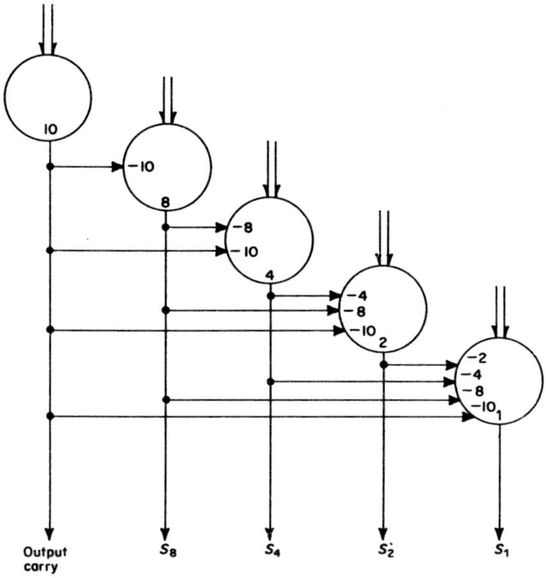

FIG. 26. Binary coded decimal adder. Primary inputs are (x_8, x_4, x_2, x_1), (y_8, y_4, y_2, y_1), and C_1, and have weights equal to their subscripts when designated by double arrows.

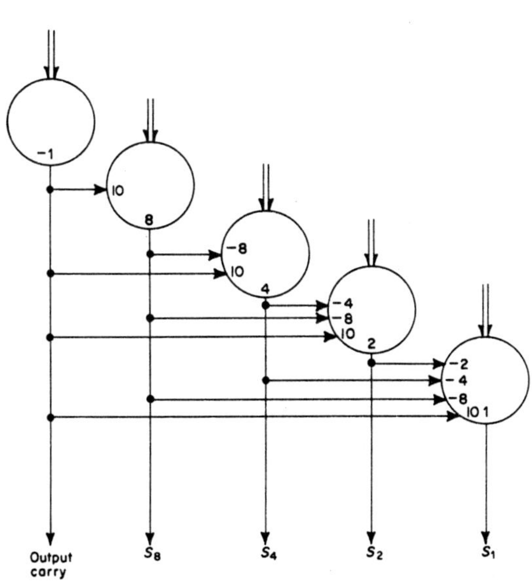

FIG. 27. Binary coded decimal subtractor. Primary inputs are (x_8, x_4, x_2, x_1), (y_8, y_4, y_2, y_1), and C_1. When designated by double arrows, the $\{x_i\}$ have weights equal to their subscripts, while the $\{y_i\}$ and C_1 have weights equal to the negative of their subscripts. The output is $(X - Y)$.

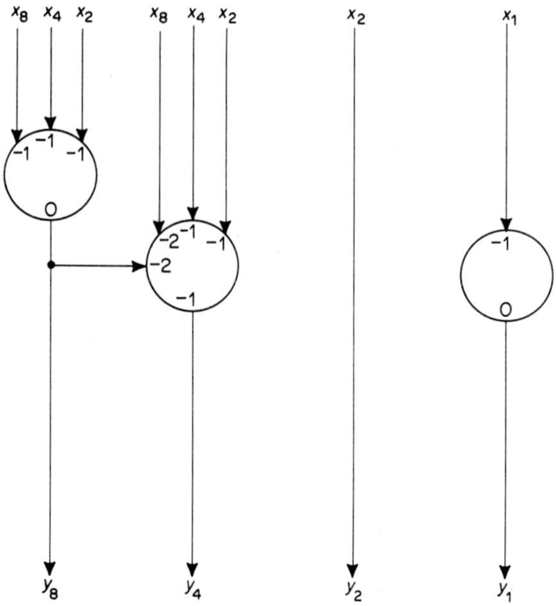

FIG. 28. Complement generator for binary coded decimal. Primary inputs are (x_8, x_4, x_2, x_1).

10.4 THRESHOLD LOGIC ELEMENT USED AS AN ERROR-CORRECTING DEVICE

Recently there has been a lot of study on the *reliability of automata*. As the system grows larger and larger, the reliability problem becomes more and more important. One method of improving reliability is the use of *redundancy techniques*. The basic principle of redundancy is the multiplication of elements, lines, modules, and even systems. Threshold logic elements play an important role in various redundancy schemes.[10, 11, 12, 13] Here we shall explain briefly the basic principles of threshold logic element used as an *error-correcting element* and as a combination of error-correcting device and computation element.

(1) Error-Correcting

Consider a function F as the output of a certain switching circuit. Let the probability of malfunctioning be p. To increase reliability it is desired to decrease the probability of malfunctioning to a tolerable value. One simple way of accomplishing this is by *triplication*, i.e. to have a triplicate of the switching circuit so that there are three independent outputs each representing the same function F. Let us designate the three outputs as F_1, F_2, and F_3, respectively. Then F_1, F_2, and F_3 are fed as inputs to a three-input majority gate, and the output of this majority gate is considered as the function F.

The function of the majority gate is $F' = F_1F_2 + F_1F_3 + F_2F_3$. Obviously the minimal-integral realization of F' is $R[F'] = (1, 1, 1; 2)$. This scheme is shown in Fig. 29.

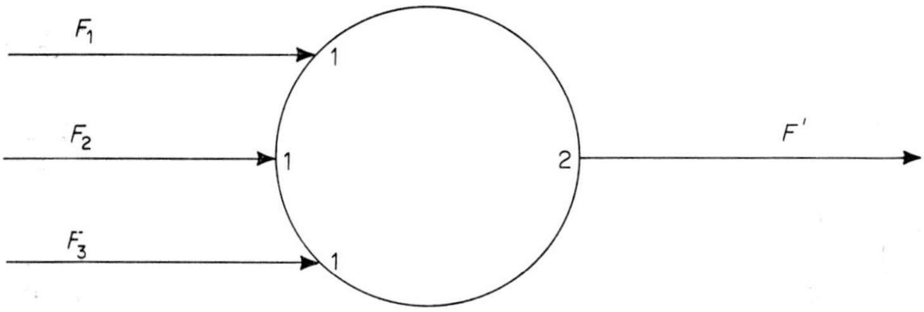

FIG. 29. Three-input majority gate.

From Fig. 29 it is seen that when F should be 1, F' is 1 if two or three of the inputs are 1 and F' is 0 if none or one of the inputs is 1. Similarly, when F should be 0, F' is 0 if two or three of the inputs are 0 and F' is 1 if none or one of the inputs is 0. In other words, F' is correct if and only if a majority, or at least two, of the three inputs are not malfunctioning. The operation of this majority gate is usually called *vote-taking*, and this gate called *vote-taker*.

Here we must make an assumption, that the vote-taker is ideal, or its probability of malfunctioning is extremely small compared with p, the probability of error of each input, for otherwise while the error of the previous circuit is corrected by the majority gate, a new error is generated in the majority gate itself, and there will be no end of correction. This assumption is justified because a three-input majority gate is relatively simple and can be made very reliable, while the previous function may be a very complex one.

With this assumption, the probability of malfunctioning of F', is

$$\begin{aligned} p' &= 3p^2(1-p)+p^3 \\ &= 3p^2-2p^3 \end{aligned} \tag{10.4–1}$$

The probability that F' is correct is

$$\begin{aligned} q &= 1-p' \\ &= 1-3p^2-2p^3 \end{aligned} \tag{10.4–2}$$

When p is small, it is obvious that $p' \ll p$ and the reliability of the circuit is improved.

(2) Error-Correcting and Computing

Since a threshold logic element can realize a function and also can correct errors in the previous stage, these two operations may be combined in one element, as was suggested by Pierce.[10] That this is possible can be explained from the "gap" point of view. Let T_u and T_l be the upper and lower boundaries of the gap. The vertices in F are mapped onto values equal to or greater than T_u, and the vertices in F are mapped onto values equal to or smaller than T_l. A malfunctioning of an input may cause a mapping to the other side of the gap, thus making an error in the output. If we can increase the gap, it may be possible that for certain errors in the inputs the mapping does not cross the gap and therefore does not result in an error in the output.

We shall explain this by using a simple example. Consider

$$F = x_1x_2 + x_1x_3 + x_2x_3$$

This function can be realized by a three-input majority gate, but when a single input is wrong, for certain input combinations the output will be wrong too. Now if instead we use a triplicate of each input and use a 5 out of 9 majority gate, then this threshold logic element is single-error correcting, in addition to its role of computing the same function. This scheme is shown in Fig. 30.

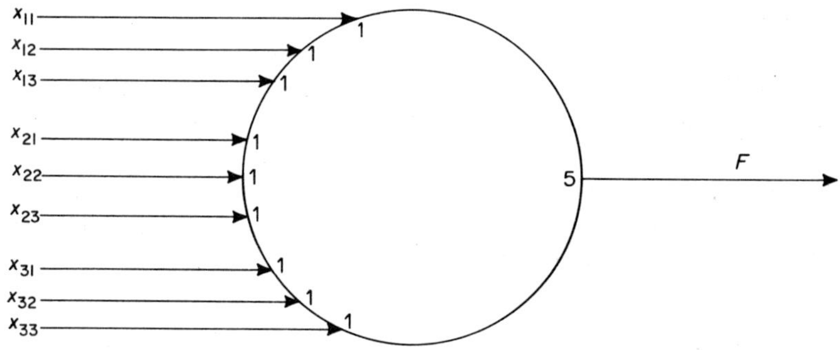

FIG. 30. Error-correcting and computing by the same threshold logic element. $F = x_1x_2 + x_1x_3 + x_2x_3$.

In Fig. 30 x_{11}, x_{12}, and x_{13} are three independent versions of x_1. Similarly x_{21}, x_{22}, x_{23} and x_{31}, x_{32}, x_{33} are independent versions of x_2 and x_3, respectively. When the inputs are all correct,

$$\sum_{i=1}^{3} \sum_{j=1}^{3} a_{ij}x_{ij} = 6 \text{ or } 9, \quad \text{for } F = 1$$

$$\sum_{i=1}^{3} \sum_{j=1}^{3} a_{ij}x_{ij} = 0 \text{ or } 3, \quad \text{for } F = 0$$

When a single input is in error,

$$\sum_{i=1}^{3} \sum_{j=1}^{3} a_{ij}x_{ij} = 5 \text{ or } 7 \text{ or } 8 \geq 5, \quad \text{for } F = 1$$

$$\sum_{i=1}^{3} \sum_{j=1}^{3} a_{ij}x_{ij} = 4 \text{ or } 2 \text{ or } 1 < 5, \quad \text{for } F = 0$$

Thus this threshold logic element performs the double functioning of correcting error in the inputs and computing the function F.

10.5 THRESHOLD LOGIC ELEMENT USED IN LEARNING SYSTEMS

One of the most important and promising fields of application of threshold logic elements is *learning systems.* Learning is a process in which a performance of the system is being improved, on the basis of past experience, by proper adjustment of some parameters of the system. This process is also known as *training.* Examples of learning systems are speech recognition, character recognition, weather forecasting, etc. Many practical learning problems can be reduced to the problem of *pattern recognition or classification.*

Pattern recognition is a process of decision making in which an input pattern is recognized as a member of a given class by a comparison of its attributes with the already known pattern of common attributes of members of that class.

Let P, the set of patterns or input events to the system, be divided into disjoint subsets, $P_1, P_2, \ldots, P_m$. Then any pattern X of P belongs to exactly one subset, i.e. $X \in P_i \subset P$, for some i.

Before a set of original objects can be recognized and classified, there is usually a device called *transducer* which transforms the original information into signals which can be accepted by the learning system. In the transducer, on each pattern X is made a sequence of numerically valued measurements $y_1, y_2, \ldots, y_n$, which may be considered as a vector in the n-dimensional space R^n. Thus, with a specified input pattern X_k is associated a vector

$$Y_k = (Y_{k1}, Y_{k2}, \ldots, Y_{kn})$$

Each of the n measurements $y_1, y_2, \ldots, y_n$ represents a property of the pattern. Numerical value of y_i, $i = 1, 2, \ldots, n$, corresponds to the amount of the ith property.

Thus the transducer can be thought of as a transformation which maps each point X of P into a point Y of the n-dimensional space R^n. The results of the measurements performed by the transducer upon X are then fed into the main part of the system which is called the *processor*. The processor decides to which class a particular X belongs, or partitions R^n into disjoint subsets $A_1, A_2, \ldots, A_m$ and makes the decision that $Y \in A_k$ if $X \in P_k$.

Besides the transducer and the processor, there is a *feedback scheme* which serves the purpose of comparing the output with the input and therefore adjusting the parameters during the training period. These three parts, transducer, processor, and feedback element, properly connected together form a self-organizing logical system called a *learning machine.*

Sometimes a certain property is either possessed or not possessed by an input pattern. In this case y_i becomes a binary variable. If a property has many different degrees of amount, then y_i should be a multi-valued variable. However, even in this case y_i can still be coded by several binary variables. Therefore there is no loss of generality in assuming that y_i is a binary variable.

If m, the number of classes, is greater than two, we can always first divide the whole set of input patterns into two disjoint subsets, each of which may be the union of several classes, and then successively divide each subset into two subsets again. Thus the multi-class problem can be reduced to a two-class problem. For a two-class problem the input patterns of one class exactly correspond to the vertices of a Boolean function F of the n variables $\{y_i\}$, and those of the other class correspond to the vertices of $\bar{F}$. If F is a threshold function, it can be separated from $\bar{F}$ by using one threshold logic element. If not, F still can be separated from $\bar{F}$ by using two or more threshold logic elements. Thus the threshold logic element finds a very interesting application in pattern classification or learning system and plays a very important role.

In pattern classification, usually the number of variables n is so large that it is impractical to test and realize a given function F by using analytical methods. Besides, it is required that the learning machine itself instead of the man will do the classification job. Therefore we should construct the machine or system in such a way that the threshold logic element can be trained to classify the patterns by successive adjustments of the parameters or weights. Such a threshold logic element is called a *trainable or adaptive threshold element*. Thus for a learning machine to classify patterns, first the machine has to go through a *training phase*, during which input patterns are presented to the machine and the weights are adjusted automatically according to a prespecified training procedure or a set of rules of adaptation. After the training has been completed, the machine is said to be in *working phase*, during which the machine will be able to classify the input events X into P_1 or P_2 by giving a correct output, say 0 for P_1 and 1 for P_2.

There have been found several sets of *rules of training or adaptation.*

The basic principle is that, if the response of the machine is correct, i.e. if $X \in P_1$ and the output of the machine is 0 or if $X \in P_2$ and the output is 1, then there will be no adjustment of weights, and that if the response of the machine is wrong, i.e. if $X \in P_1$, and the output is 1 or if $X \in P_2$ and the output is 0, then there will be an adjustment of weights. The adjustment of weights is such that if the response is 0 while it should be 1, the adjustment will be in the positive direction, i.e. the increments will be positive; whereas if the response is 1 while it should be 0, the adjustment will be in the negative direction, i.e. the increments will be negative.

There has been a very extensive study of learning systems employing threshold logic elements. For instance, Mattson,[8] Mays,[9] Glinski and Zawicki,[3] etc. have studied trainable threshold logic elements.

The crucial point in this training or adjustment of weights is that the process should be convergent if F is a threshold function. It has been proved in various ways that such a process is convergent. So the application of the trainable threshold logic element in learning systems is well established and sees a good prospect.

Because of the property of adaptability, threshold logic elements can be used in adaptive control systems as the essential adaptive element. In fact learning systems are studied as intensively from the control system point of view as from the switching theory point of view.

In fact learning is a property inherent in biological systems. Initially threshold logic element was studied as a simulation of the neuron in the nerve system. Thus a threshold logic element seems to be a very basic element in nature and will certainly play an important role in the study of cybernetics.

REFERENCES

1. Coates, C. L., and Lewis, P. M. II. (1964). Donut: A threshold gate computer, *I.E.E.E. Trans. electronic Comput.* EC**13,** 240–247.
2. Fischler, M. A. (1962). Investigations concerning the theory and synthesis of linearly separable switching functions, Ph.D. dissertation, Department of Electrical Engineering, Stanford University, U.S.A.
3. Fischler, M. A., and Poe, E. A. (1962). Threshold realization of arithmetic circuits, *I.E.E.E. Trans. electronic Comput.* EC**11,** 287–288.
4. Glinski, G. S., and Zawicki, I. J. (1964). Properties of a single "linear" threshold logic unit, *Technical Report No.* 64–14, Department of Electrical Engineering, University of Ottawa.
5. Gustafson, C. H., Haring, D. R., Susskind, A. K., and Wills-Sanford, T. G. (1965). Synthesis of counters with threshold elements, *I.E.E.E. Conference Record on Switching Circuit Theory and Logical Design,* 25–35.
6. Hadlock, F. O., and Coates, C.L. (1966). Realization of sequential machines with threshold elements, *Proc. Seventh Symp. Switching and Automata Theory,* 172–183.

7. Masters, G. M., and Mattson, R. L. (1966). Threshold logic synthesis of sequential machines, *Proc. Seventh Symp. Switching and Automata Theory,* 184–194.
8. Mattson, R. L. (1963). The analysis and synthesis of adaptive systems which use networks of threshold elements, *Technical Report No.* 1553–1, Solid-state Electronics Laboratory, Stanford Electronics Laboratories, U.S.A.
9. Mays, C. H. (1963). Adaptive threshold logic, *Technical Report No.* 1557–1, Solid-state Electronics Laboratory, Stanford Electronics Laboratories, U.S.A.
10. Pierce, W. H. (1964). Interwoven redundant logic, *J. Franklin Inst.* **277,** 55–85.
11. Pierce, W. H. (1965). Failure tolerant computer design, Academic Press, New York.
12. Sheng, C. L., and Chen, I. N. (1966). On redundancy techniques for reliable computation—a unified viewpoint from enlarging of space, *Technical Report No.* 66–13, Department of Electrical Engineering, University of Ottawa.
13. Von Neumann, J. (1956). Probabilistic logics and the synthesis of reliable organisms from unreliable components, Automata Studies, Princeton University Press.

CHAPTER 11

Threshold Logic Elements for Probability Transformation

11.1 BASIC PRINCIPLE

Probability transformer can be synthesized with a sequential circuit, as was developed by Gill.[1,2] In this chapter we shall study how threshold logic elements can be conveniently and advantageously employed for probability transformation, as was developed by Sheng.[3]

A *discrete probability transformer* is a system which converts a discrete stochastic source generating the statistically independent symbols $\alpha_1, \alpha_2, \ldots, \alpha_r$ with the probabilities $p_1, p_2, \ldots, p_r$, respectively, into another discrete stochastic source generating the statistically independent symbols $\beta_1, \beta_2, \ldots, \beta_s$ with the probabilities $q_1, q_2, \ldots, q_s$.

The idea of using a sequential circuit for probability transformation is to design a sequential machine with N states such that with the given symbols $\alpha_1, \alpha_2, \ldots, \alpha_r$ as the random inputs, eventually the machine will have a probability of $q_1 = N_1/N$ to be in a set of N_1 states, a probability of $q_2 = N_2/N$ to be in a set of N_2 states, ..., and a probability of $q_s = N_s/N$ to be in a set of N_s states, where $\sum_{i=1}^{s} N_i = N$ is the total number of states. Considering the machine as a Moore model, we can associate the s random outputs with the s sets of states, which have the desired probability distribution.

Instead of a general sequential machine with the feedback loop and with n delay units, a definite-event machine with much fewer delay units can serve this purpose of probability transformation as well. In this case the grouping of states can be very simply and conveniently done by using threshold logic elements.

First, the input alphabet $\{\alpha_1, \alpha_2, \ldots, \alpha_r\}$ is converted into a binary alphabet through a *symbol converter*. This conversion permits the usage of binary elements for the transformation logic. Let this new binary random input variable be x_0, which can be either 1 or 0, with a probability of p and $1-p$, respectively. It is preferable to split the input alphabet in such a way as to make p and $1-p$ as close to 0.5 as possible.

x_0 is fed to m delay units in cascade. Let the outputs of the delay units be $x_1, x_2, \ldots, x_m$, respectively. Then we have

$$x_1(t) = x_0(t-1)$$
$$x_2(t) = x_1(t-1) = x_0(t-2)$$
$$\cdots\cdots\cdots\cdots\cdots$$
$$x_m(t) = x_0(t-m)$$

Now the m variables, $x_1, x_2, \ldots, x_m$, each having a probability of p for being "1" and a probability of $1-p$ for being "0", form an m-tuple. There are 2^m m-tuples altogether, each m-tuple having a probability of

$$p' = p^i(1-p)^{m-i}$$

where i is the number of 1's in the m-tuple, and $m-i$ is the number of 0's in the m-tuple.

For the special case of $p = 1-p = \frac{1}{2}$, the probabilities of the 2^m m-tuples are all equal and are

$$p' = (\tfrac{1}{2})^m = 2^{-m}$$

Now if we divide the set of 2^m m-tuples into s disjoint subsets with $N_1, N_2, \ldots, N_s$ m-tuples in each subset such that

$$\sum_{i=1}^{s} N_i = N = 2^m$$

and such that

$$\sum_{j=1}^{N_i} P_{ij} = q_i \qquad (11.1\text{–}1)$$

where P_{ij} = probability of the jth m-tuple in the ith subset, then the ith subset can represent the ith output with the desired probability q_i.

This is obviously possible, because any subset of m-tuples corresponds to a Boolean function, and any Boolean function is realizable with a number of AND and OR gates.

Equation (11.1–1) can be approximated to any desired degree of accuracy by increasing m.

To ensure independence of output symbols, it is necessary to use a sampler or a pulse clock with a certain period. It is assumed that the input symbol is random. Thus the input at time t is independent of the inputs at any previous time. However, in forming the new input symbols, m delay units are used, and the input symbols are therefore dependent upon the past m input symbols. In order to obtain random output variables completely independent of past values, it is necessary to take values of either the input or the output at time intervals of at least m times the time interval of the input. For level inputs and outputs, this can be accomplished by using a sampler at the input or the output synchronized to the input time and with a sampling period $T \geq m\tau$, where τ is the period of input time interval. In the case of pulse inputs and outputs, clock pulses with a period of $T \geq m\tau$ can be applied to the input delay units.

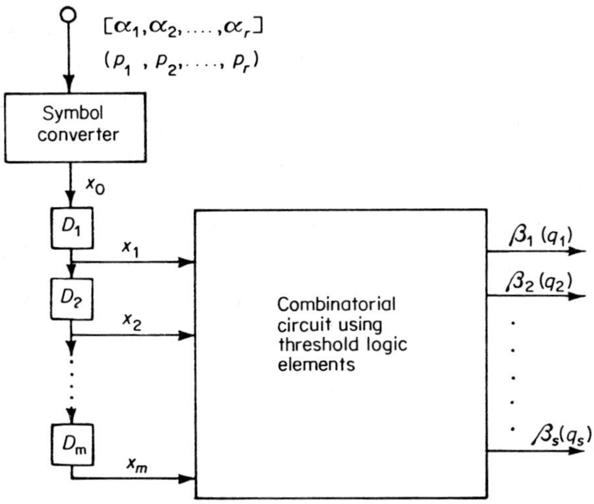

FIG. 31. Schematic diagram of a probability transformer using threshold logic elements.

Fig. 31 shows a schematic diagram of the probability transformer.

Although it is possible to realize any Boolean function with AND and OR gates, such a scheme has the disadvantage of using a large number of gates. Threshold logic elements are found to be particularly convenient and suitable for this purpose of probability transformation. The application of threshold logic elements to probability transformation essentially depends upon certain properties of isobaric functions.

THEOREM 11.1.1. The maximum number of isobaric threshold functions that can be formed from m variables for a fixed assignment of weights is equal to $2^m + 1$, including the trivial functions 1 and 0.

PROOF: Let F_i and F_{i+1} be two functions isobaric to each other and with $T_i > T_{i+1}$. Then by Theorem 4.3.3,

$$F_{i+1} \supset F_i$$

F_{i+1} must contain at least one more m-tuple than F_i, for otherwise either F_{i+1} and F_i would be identical or F_i would contain F_{i+1}. Now for m variables, there are 2^m m-tuples. Let F_1 contain 0 m-tuple, which is the trivial function 0, let F_2 contain one m-tuple, let F_3 contain two m-tuples, ..., etc. Then F_{2^m+1} contains 2^m m-tuples and is the trivial function 1. Thus there are at most $2^m + 1$ functions. Q.E.D.

THEOREM 11.1.2. The minimal assignment of integral weights to m variables to obtain $2^m + 1$ isobaric functions is 1, 2, 4, ..., 2^{m-1}.

PROOF: To obtain $2^m + 1$ functions from m variables, we must have $2^m + 1$ different threshold values, and in decreasing the threshold value from T_i to T_{i+1}, we must increase one and only one m-tuple to the function at a time. So every m-tuple must have a weight different from that of every other m-tuple, for if two m-tuples have the same weight, they will be both contained or both not contained in the function. In other words, the function containing only one of them cannot be obtained from this assignment of weights, and the total number of functions will be less than 2^m+1. Now each m-tuple has a weight equal to the sum of some i of the weights, where $i = 0, 1, \ldots, m$. The minimum integral weight is 1. Let $a_m = 1$. Then a_{m-1} cannot be 1 too, for otherwise the weights of some m-tuples, for instance, $x_1 x_2 \ldots \bar{x}_{m-1}x_m$ and $x_1x_2 \ldots x_{m-1}\bar{x}_m$, would be the same. So the minimum weight of a_{m-1} is 2. This extends to the higher weights, and the ith weight must be greater than the sum of all the previous $i-1$ weights. Thus we have

$$a_m = 1 = 2^0$$
$$a_{m-1} = a_m+1 = 2 = 2^1$$
$$a_{m-2} = a_m+a_{m-1}+1=1+2+1 = 4 = 2^2$$
....
$$a_1 = 2^{m-1}$$

Q.E.D.

COROLLARY 11.1.2–1. For the particular assignment of weights of Theorem 11.1.2, the following relation holds:

$$N+T = 2^m, \quad \text{or} \quad T = 2^m-N$$

where N = number of m-tuples contained in the function.
T = the threshold value.

PROOF: When $T = 0$, $F = 1$. In other words, the function contains all the 2^m m-tuples. Therefore Corollary 11.1.2–1 holds for $T = 0$. Now for this particular assignment of weights, the m-tuples have all the different integral weights ranging from 1 to 2^m-1, and to form all the 2^m+1 functions, T is increased by 1 at a time and therefore has all the 2^m+1 different values from 0 to 2^m. When T is increased by 1, N is decreased by 1. Therefore we have

$$T+N = 2^m$$
$$\text{for}$$
$$N = 1, 2, \ldots, 2^m.$$

Q.E.D.

11.2 REALIZATION OF PROBABILITY TRANSFORMER

We first consider the case that for $x_1, x_2, \ldots, x_m$,

$$p = 1-p = \tfrac{1}{2}$$

Then each m-tuple has a probability of $\frac{1}{2}^m$. It is required to obtain an output which is "1" with a probability of q_1. Let us express q_1 as a fraction $N_1/2^m$, or let

$$N_1 = [2^m q_1]$$

where $[2^m q_1]$ = the integer closest to $2^m q_1$.

Then N_1 represents the number of m-tuples contained in the function which has a probability equal or closest to q_1. Since each m-tuple is a random variable, we can choose any N_1 m-tuples to form the function. Here we purposely choose those N_1 m-tuples such that they form a threshold function.

Let the assignment of weights be

$$a_1 = 2^{m-1}, a_2 = 2^{m-2}, \ldots, a_{m-1} = 2, a_m = 1$$

By Corollary 11.1.2–1,

$$T_1 = 2^m - N_1$$

T_1 can be expressed as the sum of powers of 2 as

$$T_1 = c_1 2^{m-1} + c_2 2^{m-2} + \ldots + c_{m-1} 2^1 + c_m 2^0$$

where $c_1, c_2, \ldots, c_m$ = the coefficients of the powers of $2 \in \{0, 1\}$.

Then the value of c_i to be 1 or 0 represents the presence or absence of x_i in the lowest vertex (or the m-tuple with the minimum weight) in the function. For instance, in Example 1 given below, for F_1, the lowest vertex is

$$x_1 x_3 x_5 x_6 x_7$$

According to the assignment of weights of Theorem 11.1.2,

$$a_i > \sum_{j=i+1}^{m} a_j$$

So as $x_1 x_3 x_5 x_6 x_7$ is contained in the function, $x_1 x_2$ and $x_1 x_3 x_4$ must also be contained. Thus the function F_1 is easily found.

Next we are to find the function F_2 which has an output with a probability of q_2. If we give the same assignment of weight to F_2 as to either F_1, then F_1 and F_2 are isobaric to each other, and therefore either

$$F_2 \supset F_1$$

or

$$F_1 \supset F_2$$

Since q_1 and q_2 correspond to two mutually exclusive events, or two disjoint sets of m-tuples, it seems to be impossible to use the same assignment of weights. If we use a different assignment for F_2, it is still difficult to find a function completely disjoint from F_1 and yet containing the required number of m-tuples. Here we resort to the property of inhibition to threshold logic elements so that we may use the same assignment of weights. In other words we shall find a function F_2 isobaric to F_1 and having an output with a probability of q_1+q_2, and then inhibit it with F_1 so that when F_1 has an output of "1", F_2 will have an output of "0". Then the output of F_2 will have a probability q_2 which represents a mutually exclusive event from that of F_1.

The weight of inhibition should be such that the output of F_2 is "0" whenever the output of F_1 is "1". Let a_{12} denote the weight of inhibition of F_1 to F_2. Then

$$a_{12} < T_2 - \sum_{i=1}^{m} a_i$$

where T_2 = the threshold value of threshold logic element realizing F_2.

a_i = weight of input x_i to threshold logic element realizing F_2.

The same reasoning and techniques apply to $F_3, F_4, \ldots, F_5$. Thus F_3 is realized to have a probability of $q_1+q_2+q_3$. With inhibitions from F_1 and F_2, it has a probability of q_3. Similarly $F_4, \ldots, F_5$ can be realized.

Let us consider an example.

EXAMPLE 11.2–1

$p = 1-p = \frac{1}{2}$

$q_1 = 0{\cdot}32$

$q_2 = 0{\cdot}27$

$q_3 = 0{\cdot}23$

$q_4 = 0{\cdot}18$

$m = 7$

(1) Determination of threshold function F_1 with a probability of $q_1 = 0{\cdot}32$.

$$\begin{aligned} N_1 &= [2^7 \times 0{\cdot}32] \\ &= [128 \times 0{\cdot}32] \\ &= [40{\cdot}96] \\ &= 41 \\ T_1 &= 128-41 \\ &= 87 \\ &= 64+16+4+2+1 \\ &= 1\times 2^6+0\times 2^5+1\times 2^4+0\times 2^3+1\times 2^2 \\ &\quad +1\times 2^1+1\times 2^0 \end{aligned}$$

The lowest vertex is $x_1x_3x_5x_6x_7$.

$$F_1 = x_1[x_2+x_3(x_4+x_5x_6x_7)]$$

(2) Determination of threshold function F_2 with a probability of q_1+q_2.

$$\begin{aligned} q_1+q_2 &= 0{\cdot}32+0{\cdot}27 \\ &= 0{\cdot}59 \\ N_2 &= [128\times 0{\cdot}59] \\ &= [75{\cdot}52] \\ &= 76 \\ T_2 &= 128-76 \\ &= 52 \\ &= 32+16+4 \\ &= 0\times 2^6+1\times 2^5+1\times 2^4+0\times 2^3+1\times 2^2 \\ &= +0\times 2^1+0\times 2^0 \end{aligned}$$

The lowest vertex is $x_2x_3x_5$.

$$F_2 = x_1+x_2x_3(x_4+x_5)$$

The function with inhibition from F_1 is

$$F_2' = F_2\bar{F}_1$$

(3) Determination of threshold function F_3 with a probability of $q_1+q_2+q_3$

$$\begin{aligned} q_1+q_2+q_3 &= 0{\cdot}32+0{\cdot}27+0{\cdot}23 \\ &= 0{\cdot}82 \\ N_3 &= [128\times 0{\cdot}82] \\ &= [104{\cdot}96] \\ &= 105 \\ T_3 &= 128-105 \\ &= 23 \\ &= 16+4+2+1 \\ &= 0\times 2^6+0\times 2^5+1\times 2^4+0\times 2^3+1\times 2^2 \\ &\quad +1\times 2^1+1\times 2^0 \end{aligned}$$

The lowest vertex is $x_3x_5x_6x_7$.

$$F_3 = x_1+x_2+x_3(x_4+x_5x_6x_7)$$

The function with inhibition from F_1 and F_2' is

$$F_3' = F_3\bar{F}_1\bar{F}_2$$

(4) Determination of threshold function F_4 with a probability of $q_1+q_2+q_3+q_4$.

$$\begin{aligned} q_1+q_2+q_3+q_4 &= 0{\cdot}32+0{\cdot}27+0{\cdot}23+0{\cdot}18 \\ &= 1{\cdot}00 \\ F_4 &= 1 \end{aligned}$$

The function with inhibitions from F_1, F'_2 and F'_3 is

$$F'_4 = F_4\bar{F}_1\bar{F}_2\bar{F}_3 = \bar{F}_1\bar{F}'_2\bar{F}'_3 = \overline{(F_1+F'_2+F'_3)}$$

It is simply an AND gate with inputs which are the complements of F_1, F'_2, and F'_3, or is the complement of an OR gate with inputs F_1, F'_2, and F'_3.

In fact this last AND gate is not necessary. Suppose there are s output symbols $\beta_1, \beta_2, \ldots, \beta_s$, with probabilities $q_1, q_2, \ldots, q_s$, respectively. Then we need $[\log_2 s]$ binary output variables to represent them, where $[\log_2 s]$ is the smallest integer greater than or equal to $\log_2 s$. Since the output symbols are all disjoint, we can use OR gates to obtain the output variables. For instance, in Example 11.2–1, there are four output symbols. So we need two OR gates to get two output variables x_1 and x_2. As shown in Fig. 32, $z_1 z_2$ (or 11) represents β_1 with a probability 0·32, $z_1\bar{z}_2$ (or 10) represents β_2 with a probability 0·27, $\bar{z}_1 z_2$ (or 01) represents β_3 with a probability 0·23, and $\bar{z}_1\bar{z}_2$ (or 00) represents β_4 with a probability of 0·18. It is obvious that the AND gate for F_4 is not required. Thus for s output probabilities, if one threshold logic element is used for each function, only $s-1$ threshold logic elements are required, together with $[\log_2 s]$ OR gates.

11.3 METHODS OF REDUCING THE SIZES OF WEIGHTS

One disadvantage of the particular assignment of weights used above is the large magnitudes of the weights. In the above example, m was taken to be 7, and the largest weight, a_1, is equal to $2^6 = 64$. For $m = 10$, a_1 will be $2^9 = 512$. Such a large weight seems to be formidable and is certainly impractical because of errors in the threshold logic elements.

However, this disadvantage is in fact not so serious as it first appears to be. As is well known, any threshold function can be realized with many different assignments of weights. The particular assignment of weights used is minimal for the generation of 2^m+1 isobaric functions, but is not minimal for any particular function. The purpose of using this assignment of weights in the beginning is to find out the threshold function having the required number of m-tuples and therefore the required probability. Once the function is determined, this particular assignment of weights is no longer required, and it is usually possible to readjust the weights to much smaller magnitudes. For instance, in Example 11.2–1, F_1 can be realized with the following minimal assignment of integral weights.

$$a_1 = 11,\ a_2 = 7,\ a_3 = 4,\ a_4 = 3,\ a_5 = a_6 = a_7 = 1,\ T = 18$$

For F_2, $a_1 = 5, a_2 = a_3 = 2, a_4 = a_5 = 1, a_6 = a_7 = 0, T = 5$

And for F_3, $a_1 = a_2 = 7$, $a_3 = 4$, $a_4 = 3$, $a_5 = a_6 = a_7 = 1$, $T = 7$

The realization of Example 11.2–1 is shown in Fig. 32.

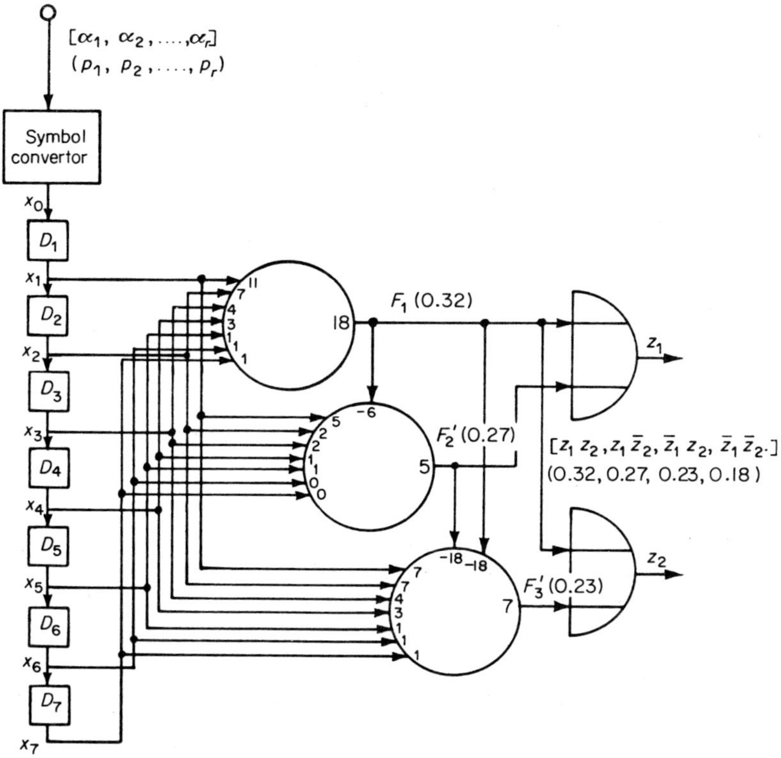

FIG. 32. Realization of Example 11.2–1.

If after readjustment the weights are still considered to be too large for physical realization, each threshold logic element may be decomposed into two or more threshold logic elements with less input variables of smaller weights. The application of decomposition to our case is quite simple, and can be easily done without the use of any systematic method. Consider Example 11.2–1 again.

$$\begin{aligned} F_1 &= x_1[x_2 + x_3(x_4 + x_5 x_6 x_7)] \\ &= x_1(x_2 + x_3 x_4) + x_1 x_3 x_5 x_6 x_7 \end{aligned}$$

Thus F_1 can be realized with two threshold logic elements, as shown in Fig. 33a.

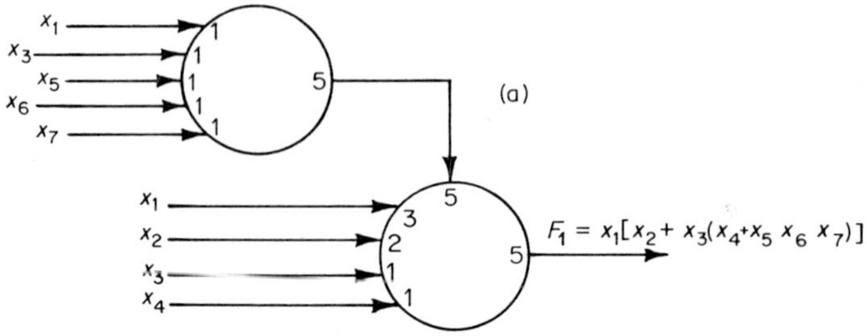

F_2 is fairly simple and does not need to be decomposed.
F_3 can be decomposed as

$$
\begin{aligned}
F_3 &= x_1 + x_2 + x_3(x_4 + x_5 x_6 x_7) \\
&= (x_1 + x_2 + x_3 x_4) + x_3 x_5 x_6 x_7
\end{aligned}
$$

The realization of F_3 with two threshold logic elements is shown in Fig. 33b.

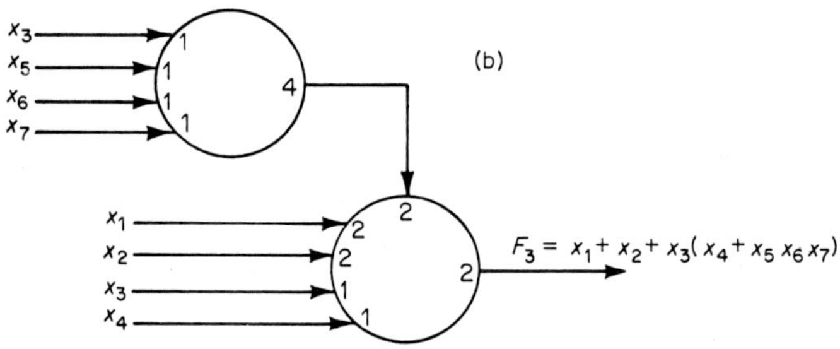

FIG. 33. Examples of decomposition of threshold logic elements. (a) Realization of F_1 of Example 11.2–1 with two threshold logic elements. (b) Realization of F_3 of Example 11.2–1 with two threshold logic elements.

11.4 GENERAL METHOD OF REALIZATION

So far we have considered only the special case for $p = 1-p = \frac{1}{2}$. For the general case, although for m variables there are still 2^m m-tuples, the probability of each m-tuple is no longer the same, but is different depending upon how many 1's it has. So the problem is now to find a threshold function such that the probability of the output to be "1" is equal to a specified probability, say q_1.

Owing to the particular way of assigning the weights, the threshold functions generated are all of the same form. If suitable brackets and parentheses are used, each variable appears only once in the expression, and the variables appear in the order as the ascending order of the subscripts of the variables. For instance, F_1 in Example 11.2–1 is

$$F_1 = x_1[x_2+x_3(x_4+x_5x_6x_7)]$$

From this expression, the number of m-tuples contained in it can easily be found.

This can be more clearly seen if F_1 is changed to the following form:

$$F_1 = x_1[x_2+\bar{x}_2x_3(x_4+\bar{x}_4x_5x_6x_7)] \qquad (11.4\text{–}1)$$

A term like x_1 contains half of the 2^m m-tuples, a term like x_1x_2 contains 1/4 of the m-tuples, etc. With the complemented variables introduced, all the terms become disjoint, and the total number of m-tuples of the function is now equal to the sum of the m-tuples contained in each term. So the total number of m-tuples contained in F_1 is equal to 2^m multiplied by the above expression with each variable, complemented or uncomplemented, replaced by $\frac{1}{2}$. Thus

$$\begin{aligned} N_1 &= 128\times+\tfrac{1}{2}[\tfrac{1}{2}+\tfrac{1}{2}\times\tfrac{1}{2}(\tfrac{1}{2}+\tfrac{1}{2}\times\tfrac{1}{2}\times\tfrac{1}{2}\times\tfrac{1}{2})] \\ &= 128\times\frac{41}{128} \\ &= 41 \end{aligned}$$

This checks with Example 11.2–1.

For the special case of $p = 1-p = \frac{1}{2}$, since $q_1 = \dfrac{N_1}{2^m}$, the probability of F_1 is simply the expression of F_1 in the form of Equation 11.4–2 with each variable replaced by $\frac{1}{2}$. For the general case, however, this is not true, and the expression for probability should be slightly modified. Each uncomplemented variable should be replaced by p, and each complemented variable should be replaced by $(1-p)$. Thus the probability of F_1 becomes

$$q_1 \simeq p\{p+(1-p)p[p+(1-p)p^3]\}$$

Now the problem reduces to: given p, $(1-p)$, and q_1 (m depends on error tolerance and may be assumed to be given too), to find the threshold function F_1.

The procedure of doing this is as follows:

Step 1. If $q_1 < p$, divide q_1 by p. If the quotient is still less than p, divide it by p again. Repeat this until the quotient is greater than or equal to p. Denote the number of times of division by d_1. Thus q_1 is already divided by p^{d_1} to become greater than or equal to p. Then subtract p from the last quotient, which is greater or equal to p, to obtain the difference D. Go to Step 3.

Step 2. If $q_1 \geq p$, subtract p from q_1 to obtain the difference D. Go to Step 3.

Step 3. If the difference in Step 1 or 2 is zero, the procedure is already completed. If the difference D is greater than zero, divide D by $(1-p)$. Get the quotient q_1'. Then treat q_1' as q_1 and repeat Steps 1 or 2. In Step 1 the new number of times of division is denoted by d_2.

Step 4. Repeat Steps 1 through 3 until the mixed highest power of p and $(1-p)$ (in the last term) is equal to m.

Step 5. Arrange the p's and $(1-p)$'s in the following form, which is equal or closest to q_1 for the given m.

$$q_1 \simeq p^{d_1}\{p+(1-p)p^{d_2}[p+(1-p)p^{d_3}(p+\ldots)]\} \qquad (11.4\text{–}2)$$

Step 6. Form F_1 according to Equation 11.4–2 with each p replaced by an uncomplemented variable and each $(1-p)$ replaced by a complemented variable, starting from x_1 with each uncomplemented variable appearing exactly once in the ascending order of the subscript and with each complemented variable having the same subscript as the preceding uncomplemented variable.

Let us consider the above example again, but with different p and $(1-p)$.

EXAMPLE 11.4–1

$$p = 0{\cdot}7$$
$$1-p = 0{\cdot}3$$
$$q_1 = 0{\cdot}32$$
$$q_2 = 0{\cdot}27$$
$$q_3 = 0{\cdot}23$$
$$q_4 = 0{\cdot}18$$
$$m = 7$$

(1) Determination of F_1.

$$q_1 = 0{\cdot}32 < 0{\cdot}7$$
$$\frac{0{\cdot}32}{0{\cdot}7} = 0{\cdot}457 < 0{\cdot}7$$
$$\frac{0{\cdot}457}{0{\cdot}7} = 0{\cdot}653 < 0{\cdot}7$$
$$\frac{0{\cdot}653}{0{\cdot}7} = 0{\cdot}947 > 0{\cdot}7$$
$$d_1 = 3$$
$$0{\cdot}947 - 0{\cdot}7 = 0{\cdot}247$$

$$\frac{0{\cdot}247}{0{\cdot}3} = 0{\cdot}823 > 0{\cdot}7$$
$$d_2 = 0$$
$$0{\cdot}823 - 0{\cdot}7 = 0{\cdot}125$$
$$\frac{0{\cdot}123}{0{\cdot}3} = 0{\cdot}41 < 0{\cdot}7$$
$$\frac{0{\cdot}41}{0{\cdot}7} = 0{\cdot}586 < 0{\cdot}7$$
$$\frac{0{\cdot}586}{0{\cdot}7} = 0{\cdot}837 > 0{\cdot}7$$
$$d_3 = 2$$

The mixed highest power is already 7. So we stop here.

$$\begin{aligned} q_1 &= 0{\cdot}32 \\ &\simeq 0{\cdot}7^3[0{\cdot}7 + 0{\cdot}3(0{\cdot}7 + 0{\cdot}3 \times 0{\cdot}7^2)] \\ &= 0{\cdot}327 \\ F_1 &= x_1 x_2 x_3[x_4 + \bar{x}_4(x_5 + \bar{x}_5 x_6 x_7)] \\ &= x_1 x_2 x_3(x_4 + x_5 + x_6 x_7) \end{aligned}$$

The minimal assignment of integral weights is

$$a_1 = a_2 = a_3 = 5,\ a_4 = a_5 = 2,\ a_6 = a_7 = 1,\ T = 17$$

(2) Determination of F_2.

$$q_1 + q_2 = 0{\cdot}32 + 0{\cdot}27 = 0{\cdot}59 < 0{\cdot}7$$
$$\frac{0{\cdot}59}{0{\cdot}7} = 0{\cdot}843 > 0{\cdot}7$$
$$d_1 = 1$$
$$0{\cdot}843 - 0{\cdot}7 = 0{\cdot}143$$
$$\frac{0{\cdot}143}{0{\cdot}3} = 0{\cdot}477 < 0{\cdot}7$$
$$\frac{0{\cdot}477}{0{\cdot}7} = 0{\cdot}681 < 0{\cdot}7$$
$$\frac{0{\cdot}681}{0{\cdot}7} = 0{\cdot}973 > 0{\cdot}7$$
$$d_2 = 2$$
$$0{\cdot}973 - 0{\cdot}7 = 0{\cdot}273$$
$$\frac{0{\cdot}273}{0{\cdot}3} = 0{\cdot}91 > 0{\cdot}7$$
$$d_3 = 0$$

$$0{\cdot}91 - 0{\cdot}7 = 0{\cdot}21$$

$$\frac{0{\cdot}21}{0{\cdot}3} = 0{\cdot}7 = 0{\cdot}7$$

$$d_4 = 0$$

$$0{\cdot}7 - 0{\cdot}7 = 0$$

The difference is equal to zero. So we stop here.

$$\begin{aligned} q_1 + q_2 &= 0{\cdot}59 \\ &\equiv 0{\cdot}7\{0{\cdot}7 + 0{\cdot}3 \times 0{\cdot}7^2[0{\cdot}7 + 0{\cdot}3(0{\cdot}7 + 0{\cdot}3 \times 0{\cdot}7)]\} \\ &= 0{\cdot}590 \\ F_2 &= x_1\{x_2 + \bar{x}_2 x_3 x_4[x_5 + \bar{x}_5(x_6 + \bar{x}_6 x_7)]\} \\ &= x_1[x_2 + x_3 x_4(x_5 + x_6 + x_7)] \\ F'_2 &= F_2 \bar{F}_1 \\ a_1 &= 10,\ a_2 = 7,\ a_3 = a_4 = 3,\ a_5 = a_6 = a_7 = 1 \\ T &= 17 \end{aligned}$$

(3) Determination of F_3.

$$q_1 + q_2 + q_3 = 0{\cdot}32 + 0{\cdot}27 + 0{\cdot}23 = 0{\cdot}82 > 0{\cdot}7$$

$$d_1 = 0$$

$$0{\cdot}82 - 0{\cdot}7 = 0{\cdot}12$$

$$\frac{0{\cdot}12}{0{\cdot}3} = 0{\cdot}40 < 0{\cdot}7$$

$$\frac{0{\cdot}40}{0{\cdot}7} = 0{\cdot}571 < 0{\cdot}7$$

$$\frac{0{\cdot}571}{0{\cdot}7} = 0{\cdot}816 > 0{\cdot}7$$

$$d_2 = 2$$

$$0{\cdot}816 - 0{\cdot}7 = 0{\cdot}116$$

$$\frac{0{\cdot}116}{0{\cdot}3} = 0{\cdot}387 < 0{\cdot}7$$

$$\frac{0{\cdot}387}{0{\cdot}7} = 0{\cdot}553 < 0{\cdot}7$$

$$\frac{0{\cdot}553}{0{\cdot}7} = 0{\cdot}790 > 0{\cdot}7$$

$$d_3 = 2$$

$$0{\cdot}790 - 0{\cdot}7 = 0{\cdot}090$$

$$\frac{0{\cdot}090}{0{\cdot}3} = 0{\cdot}300 < 0{\cdot}7$$

The mixed highest power of p and $(1-p)$ is already seven. So we stop here.

$$\begin{aligned} q_1+q_2+q_3 &= 0{\cdot}82 \\ &\simeq 0{\cdot}7+0{\cdot}3\times 0{\cdot}7^2[0{\cdot}7+0{\cdot}3\times 0{\cdot}7^2(0{\cdot}7+0{\cdot}3)] \\ &= 0{\cdot}7+0{\cdot}3\times 0{\cdot}7^2(0{\cdot}7+0{\cdot}3\times 0{\cdot}7^2) \\ &= 0{\cdot}825 \end{aligned}$$

$$\begin{aligned} F_3 &= x_1+\bar{x}_1x_2x_3(x_4+\bar{x}_4x_5x_6) \\ &= x_1+x_2x_3(x_4+x_5x_6) \\ F'_3 &= F_3\bar{F}_1\bar{F}'_2 \\ a_1 &= 8,\ a_2 = a_3 = 3,\ a_4 = 2,\ a_5 = a_6 = 1,\ a_7 = 0 \\ T &= 8 \end{aligned}$$

(4) Determination of F_4.

$$\begin{aligned} q_1+q_2+q_3+q_4 &= 0{\cdot}32+0{\cdot}27+0{\cdot}23+0{\cdot}18 = 1{\cdot}00 \\ F_4 &= 1 \\ F'_4 &= F_4\bar{F}_1\bar{F}'_2\bar{F}'_3 \\ &= \bar{F}_1\bar{F}'_2\bar{F}'_3 \end{aligned}$$

The error in using threshold logic elements as a probability transformer depends on m. The larger m, the smaller the error. For $p = 1-p = \frac{1}{2}$, since N_i is the closest integer to $2^m q_2$, the error bound is half of the probability of the m-tuple, or $\frac{1}{2}\times\frac{1}{2^m} = \frac{1}{2^{m+1}}$. For some probabilities, such as q_2 and q_3 in the above example, the probability is the difference between two probabilities each with an error bound of $1/(2^{m+1})$. So the error bound is twice of this, or

$$\epsilon = 2\times\frac{1}{2^{m+1}} = \frac{1}{2^m}$$

In the case of $m = 7$,

$$\epsilon = \frac{1}{2^7} = \frac{1}{128} = 0{\cdot}0078 = 0{\cdot}78\%$$

For $p \neq 1-p \neq \frac{1}{2}$, each m-tuple has a different probability. The error bound is more difficult to calculate, and is roughly estimated that for the same m, the error bound for $p \neq 1-p \neq \frac{1}{2}$ is slightly larger than that for $p = 1-p = \frac{1}{2}$, as indicated in the above examples.

REFERENCES

1. Gill, A. (1962). Synthesis of probability transformer, *J. Franklin Inst.* **274,** 1–19.
2. Gill, A. (1963). On a weight distribution problem with application to the design of stochastic generators, *J. Ass. comput. Mach.* **10,** 110–121.
3. Sheng, C. L. (1965). Threshold logic elements used as a probability transformer, *J. Ass. comput. Mach.* **12,** 262–276.

Subject Index

R

S

T

131069WRL